Professional Fiber Optic Installation v10 + FOA Certification

PowerPoint Slides

Version 7-30-2024

Eric R. Pearson,

CFOS/S/C/T/H/O/I

Disclaimer

All instructions contained herein are believed to produce theproper results when followed exactly with appropriate equipment. However, these instructions are not guaranteed for all situations.

Notice To The Reader

The publisher does not warrant or guarantee any of the products described herein or perform any independent analysis in connection with any of the product information contained herein. Publisher does not assume, and expressly disclaims, any obligation to obtain and include information other that provided to it by the manufacturer.

The reader is specifically warned to consider and adopt any and all safety precautions that might be indicated by the activities herein and to avoid any and all potential hazards. By following the instructions contained herein, the reader knowingly and willingly assumes all risks in connection with such instructions.

The publisher makes no representation or warranties of any kind, included but not limited to, the warranties of fitness for particular purpose or merchantability, nor are any such representations implied with respect to the material set forth herein, and the publisher takes no responsibility with respect to such material. The publisher shall not be liable for any special, consequential, or exemplary damages resulting, in whole or part, from the readers' use of, or reliance upon, this material.

The procedures provided herein are believed to be accurate and to result in low installation cost and in high reliability. However, there is a possibility that these procedures may be unsuitable for specific products or in specific situations. Because of this possibility, the installer should review, and follow, the instructions provided by product manufacturers. The instructions contained herein are not meant to imply that conflicting instructions from manufacturers are in error.

Trademark Notice

Publishedby

PearsonTechnologiesInc.
4671 HickoryBend Drive
AcworthGA30102
770-490-9991
www.ptnowire.com
fiberguru311@gmail.com

Version Date: 7-30-23

File: POFI-PowerPoint-V3-7-30-24.pdf

Printedin the United States of America.

10987654321

979885454379

Pearson Technologies Inc.

4671 Hickory Bend Dr Acworth GA 30102 770-490-9991 www.ptnowire.com fiberguru@ptnowire.com

July 30, 2024

About Professional Fiber Optic Installation + CFOT Certification

Program Focii

This program has two foci. The first focus is to provide you with the knowledge and practice necessary to be successful in installation activities after the program. The second focus is to provide the information in a format that enables passing the Fiber Optic Association (FOA) Certified Fiber Optic Technician (CFOT) certification examination. The objective of this PowerPoint book is to enable passing the CFOT examination.

The first focus includes presentations on and hands-on practice in the activities required for successful fiber optic installation. These activities include: cable end preparation, fusion splicing, connector installation, pigtail splicing, insertion last testing, OTDR testing and interpretation, connector microscopic inspection, use of the VFL, and calculation of acceptance values for insertion loss and OTDR test results.

In deliveries of this program focused on developing advanced level skills, these activities include attachment of cables to an enclosure, multiple fusion splices, placement of fiber in splice trays, and dressing the enclosure.

Certification Process

This program format enables passing the CFOT certification examination. It does so in five ways:

First, red print in the presentation slides indicates information that is important to know.

Second, if highlighting the red information enables focusing on studying that information.

Third, this package includes FOA practice quizzes. Answering these quizzes outside of the training program, enables preparation for the examination.

Fourth, the class review of the answers to the review quizzes provides the correct quiz answers. Studying answers after the class review enables focusing on the potential examination questions.

4671 Hickory Bend Dr Acworth GA 30102 770-490-9991 www.ptnowire.com fiberguru@ptnowire.com

Fifth, prior to the examination, there will be time for review.

An alternative approach for preparation is taking the practice quizzes online. The master list to the lessons and quizzes is at:

https://fiberu.org/basic/index.html

The individual quizzes are at:

https://www.thefoa.org/tech/ref/basic/quiz/jargon/jargon.htm

https://www.thefoa.org/tech/ref/basic/quiz/link/link.htm

https://www.thefoa.org/tech/ref/basic/quiz/basic/basic.htm

https://www.thefoa.org/tech/ref/basic/quiz/networks/networks.htm

https://www.thefoa.org/tech/ref/basic/quiz/fiber/fiber.htm

https://www.thefoa.org/tech/ref/basic/quiz/cables/cables.htm

https://www.thefoa.org/tech/ref/basic/quiz/term/term.htm

https://www.thefoa.org/tech/ref/basic/quiz/design/design.htm

https://fiberu.org/Design/Quizzes/LP4quiz.htm

https://www.thefoa.org/tech/ref/basic/quiz/install/install.htm

https://www.thefoa.org/tech/ref/basic/quiz/test/test.htm

The certification examination questions are chosen from the questions in the review quizzes. 71 questions are randomly chosen from the ~141 review questions. A passing grade is 70%, or 71 correct answers.

Acknowledgements

Pearson Technologies Inc. thanks the Mr. Jim Hayes of the Fiber Optic Association for permission to include the practice quizzes in this book. In addition, Pearson Technologies appreciates Mr. Hayes many years of support.

About Pearson Technologies Inc.

For information on fiber training programs and technical support for fiber lawsuits, visit: https://www.ptnowire.com/

Eric R. Pearson, CFOS/T/C/S/H/O/I

Pearson Technologies Inc.

4671 Hickory Bend Dr Acworth GA 30102 770-490-9991 www.ptnowire.com fiberguru@ptnowire.com

FOA Master Instructor
BICSI Master Instructor

Author of:
Professional Fiber Optic Installation: The Essentials for Success, v10
Mastering The OTDR: Trace Acquisition And Interpretation
The Installer's Guide to Fiber Optic Communications- Essential Understanding of Fiber Transmission
The Installer's Guide to Fiber Optic Testing-Essential Understanding of Principles and Procedures
Fiber Optic Communications For Beginners: The Basics
Successful Fiber Optic Installation: A Rapid Start Guide
Mastering Fiber Optic Connector Installation: A Guide To Low Loss, Low Cost, And High Reliability
Mastering Fiber Optic Network Design: The Essentials
PON Design And Installation, v 7.0

www.fiberopticlawsuits.com

/ fiberguru311@gmail.com

43 Years Of Superior Fiber Optic Training And Consulting

File: 0 About this Program.docx

43 Years Of Superior Fiber Optic Training And Consulting

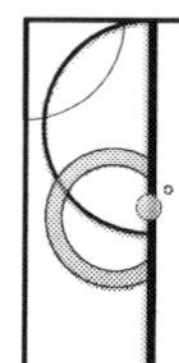

Professional Fiber Optic Installation With CFOT Certification

The Essentials For Success

Developed And Delivered By
Eric R. Pearson, CFOS
FOA Master Instructor
BICSI Master Instructor
Pearson Technologies Inc.

1

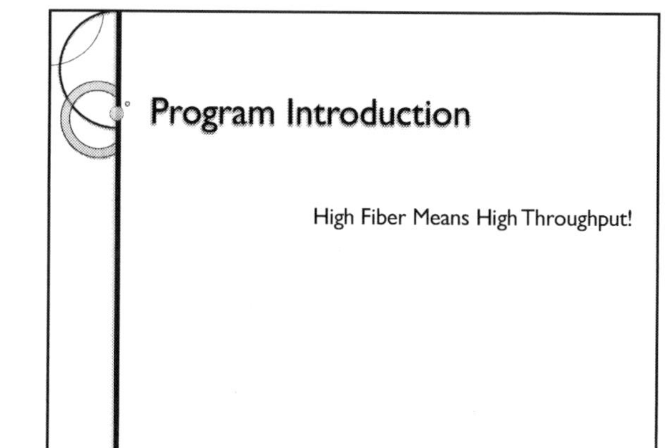

2

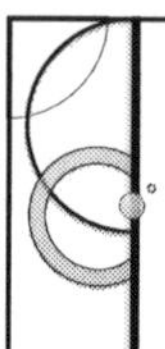

Program Objectives

One goal: be successful!

3

Program Objectives

- Successful means
  - Low installation cost
  - Low power loss
  - High reliability
- Pass basic certification examination, Certified Fiber Optic Technician (CFOT)

© Pearson Technologies Introduction 4

4

Success Through

- Learning the basics of installation
- Learning the hands-on techniques for success
- Developing basic connector, cable, and splice installation, inspection & testing skills

© Pearson Technologies Introduction 5

5

Approach

- Lecture
- Hands-on practice
  - To reinforce the lecture materials
- Repetitions
  - To 'climb the learning curve'
- For certifications
  - Review, review, review
  - Take online practice quizzes (see practice quizzes at end of specific Lesson Plans at: https://fiberu.org/basic/index.html)
  - Preparation outside of course: review
    - Professional Fiber Optic Installation, v10 + CFOT Certification PowerPoint and
    - Professional Fiber Optic Installation, v10

© Pearson Technologies Introduction 6

6

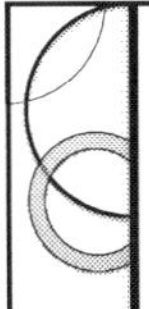

Lecture based on

- Professional Fiber Optic Installation, v10
  - Presentation Chapters 1-6, 8, 10, 11, 14-15, 18-20
  - Hands-On Chapters 14-15, 19-20, 21, 24, 26, 27, 28
- FOA Reference Guide to Fiber Optics

© Pearson Technologies Introduction 7

7

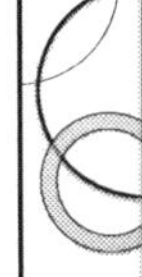

Hands-On Practice 1

- Handle fiber without breakage
  - Learn when to be careful and when to be very careful
- Prepare cable ends
  - Loose tube
  - Tight tube
- Install connectors
  - 2 multimode Unicam- no polish*
  - 2 singlemode SOCs or 2 singlemode pigtails

© Pearson Technologies Introduction 8

8

Hands-On Practice 2

- Splicing
  - Make and test fusion splice
    - Mid span
    - Pigtail
- Test insertion loss
  - Learn how to make and interpret test results
- Test with OTDR
  - Learn to interpret traces to find locations of high loss

© Pearson Technologies Introduction 9

9

Hands-On Practice 3

- Use VFL
  - Learn how to evaluate high loss of Unicams
- Inspect connectors
  - Learn how to inspect and evaluate microscopic appearance of all connectors
- Calculate acceptance values
  - For high reliability
- Troubleshooting
  - Learn to recognize and locate conditions of reduced reliability and high loss

*Unicam® connectors not included in deliveries with Advanced Splicing.

© Pearson Technologies Introduction 10

10

Hands-On Activities

- Chapters 14-15, 19-26 include instructions for hands-on activities

© Pearson Technologies Introduction 11

11

Certification

- Certification is a Fiber Optic Association (FOA) certification
- FOA has issued more than 92,225 certifications worldwide (as of 7/2023)
- Certification requires passing an examination
- Basic certification: Certified Fiber Optic Technician, CFOT
- Certification examination is on-line: bring your computer or tablet on Friday

© Pearson Technologies Introduction 12

12

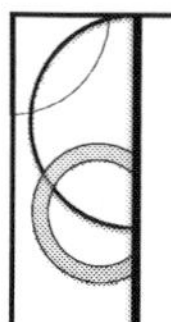

Red Text

- Red text indicates important information
  - It may be important because the instructor thinks you need this understanding to be successful during installation
  - It may be important because the Fiber Optic Association thinks you need this understanding to pass the CFOT certification examination
- Suggestion: highlight red text in your PowerPoint notes.
  - You need highlight only the start of the red text, not the entire red passage.
- Prepare for the examination by studying the highlighted information
  - It is possible that every examination answer is in red

© Pearson Technologies Introduction 13

13

Examination Preparation Format

- First, red print in the presentation slides indicates information that is important to know.
- Second, highlighting red information enables focus on studying that information in preparation for the examination.
- Third, this package includes FOA practice quizzes. By answering these quizzes outside of the training program, you prepare yourself for the examination.
- Fourth, the class review of the answers to the review quizzes provides the correct quiz answers. By studying answers after the class review, you focus on the potential examination questions.
- Fifth, take the online quizzes and review the answers
- Sixth, there will be time for review prior to the examination.

© Pearson Technologies Introduction 14

14

Review Questions 1

- When answering the review questions, note your answer and the slide(s) within which you found the answer
  - Note the slide in the format section number-slide number.
- During class review, share the locations with the class
- The quizzes appear at the end of each PowerPoint slide section
- The quizzes do not parallel the sections completely.
  - If you cannot answer the questions, do not be concerned. Subsequent sections will enable you to do so.

© Pearson Technologies Introduction 15

15

The Online Practice Quizzes are at:

- https://fiberu.org/basic/index.html
- https://www.thefoa.org/tech/ref/basic/quiz/jargon/jargon.htm
- https://www.thefoa.org/tech/ref/basic/quiz/link/link.htm
- https://www.thefoa.org/tech/ref/basic/quiz/basic/basic.htm
- https://www.thefoa.org/tech/ref/basic/quiz/networks/networks.htm
- https://www.thefoa.org/tech/ref/basic/quiz/fiber/fiber.htm
- https://www.thefoa.org/tech/ref/basic/quiz/cables/cables.htm
- https://www.thefoa.org/tech/ref/basic/quiz/term/term.htm
- https://www.thefoa.org/tech/ref/basic/quiz/design/design.htm
- https://fiberu.org/Design/Quizzes/LP4quiz.htm
- https://www.thefoa.org/tech/ref/basic/quiz/install/install.htm
- https://www.thefoa.org/tech/ref/basic/quiz/test/test.htm

© Pearson Technologies Introduction 16

16

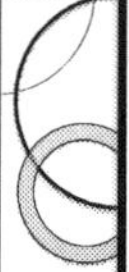

About the Examinations

- The instructor helped write the examinations
  - The instructor was a Director of the Fiber Optic Association for 12 years
- Answer the questions in the review quizzes that are included in the PowerPoint book.
- We will review the answers as a class.
- Because your have answered these questions, you will be able to pass the CFOT examination
- The examination has 71 questions. 70% (50 correct answers) needed to pass.
- Preparation outside of class will help you pass.

© Pearson Technologies Introduction 17

17

Daily Schedule

- For 5 Day Presentation
- Program: 0800-1700
- Lunch: 1130-1230
- Breaks: ~1000-1015, ~1400-1415

© Pearson Technologies Introduction 18

18

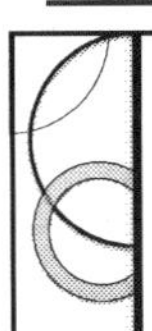

Your Responsibilities

- Listen for understanding
- Ask questions to achieve understanding
- Take notes if necessary
  - You may mark the book- it's yours
  - Almost everything I'll say is in the PowerPoint book and Professional Fiber Optic Installation, v10
  - Minimum note taking required
- Use the language in class
- Do not fear errors
- Be on time

© Pearson Technologies Introduction 19

19

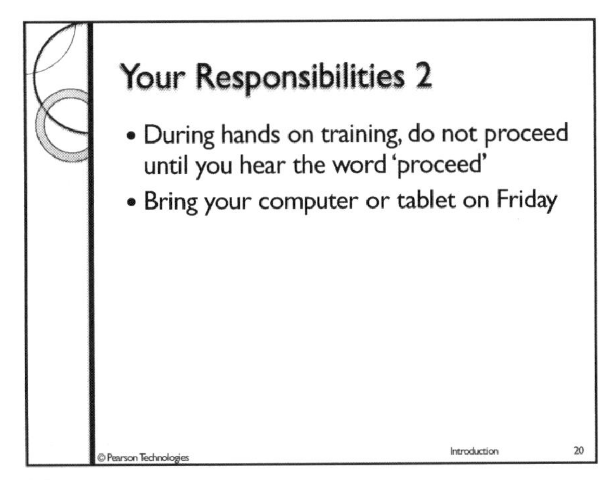

20

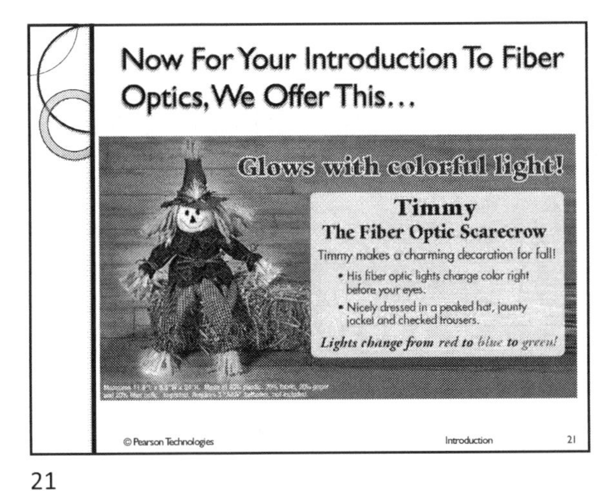

21

End Of Introduction

Questions?
Comments?
Observations?

22

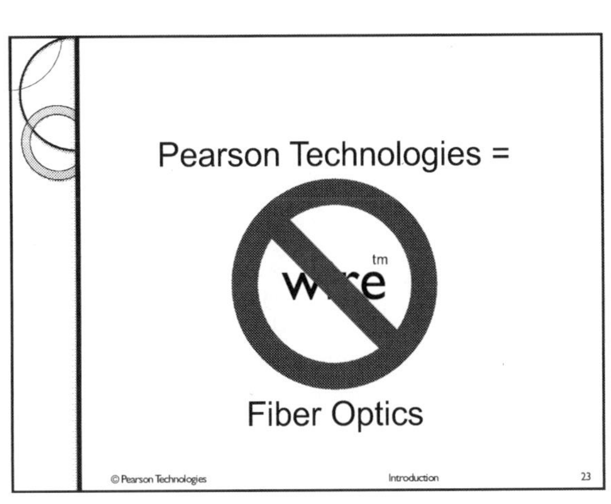

23

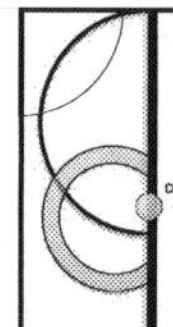

Professional Fiber Optic Installation

The Essentials For Success

Developed And Delivered By
Eric R. Pearson, CFOS
FOA Master Instructor
BICSI Master Instructor
Pearson Technologies Inc.

1

Program Objectives

One goal: be successful!

2

Program Objectives

- Success means
  - Low installation cost
  - Low power loss
  - High reliability
- Learn the basics of installation
- Learn the hands-on techniques for success
- Develop basic connector, cable, and splice installation, inspection & testing skills
- Pass basic certification examination, Certified Fiber Optic Technician (CFOT)

3

A Light Overview

Chapter 2

Where Nothing Is Heavy
When Learning About Light!

4

The Language Of Light

- Concepts
- Properties
- Behavior
- Numbers
- Significance of the different types of fiber (Chapter 3)

5

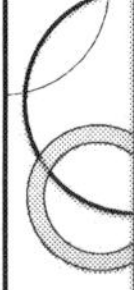

Concepts For Description

- Rays
  - Light travels in straight line in multimode fibers
- Particles
  - Light reflects/bounces in multimode fibers
- Waves, or energy fields
  - Light energy occupies a volume in singlemode fibers

6

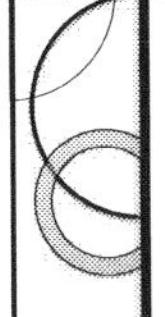

The Language Of Light

- Concepts
- Properties
- Behavior
- Numbers
- Significance of the different types of fiber (Chapter 3)

7

Light Properties (2-2)

- Wavelength
- Spectral width
- Speed
- Power
- Volume
- Pulse Width
- Polarization (Advanced Concepts)
- Phase (Advanced Concepts)
- Wavelength non-interference

8

Wavelength

- When we think about light, we think of color
- The technical term for color is 'wavelength'
- Light has a periodic, or wave-like, nature
- The period of repetition is its wavelength
- Example
  - The distance from peak to peak is the 'wavelength' of the waves in the water
- Fiber wavelengths are invisible, infrared (IR) light, with potentially dangerous power levels

9

Significance of Wavelength

- The wavelength determines
  - Dispersion in the fiber
    - Dispersion determines signal accuracy and maximum distance of transmission
  - Power loss in the fiber
    - Power loss determines maximum distance of transmission

10

Wavelength And Dispersion

- Even if there is enough power at the receiver, when dispersion is excessive, the receiver cannot convert the received power to an electrical signal identical to the input electrical signal

11

Wavelength And Power Loss

- If power loss is excessive, the receiver cannot convert the received power to an electrical signal identical to the input electrical signal

12

Measuring Wavelength (2-1)

- The wavelength, λ, is the measure of the distance between the peaks, or between the troughs, in nanometers (nm)
- Wavelength ranges from 780 nm to 1625 nm
- Most data networks use wavelengths between 850 and 1550 nm
  - Multimode wavelengths are 850nm and 1300nm
  - Singlemode wavelengths are 1310nm, 1550nm, and wavelengths in between

 1- A Light Overview 13

13

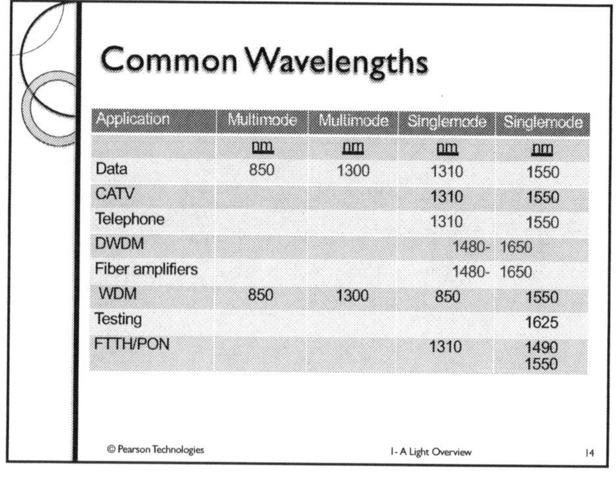

Common Wavelengths

Application	Multimode	Multimode	Singlemode	Singlemode
	nm	nm	nm	nm
Data	850	1300	1310	1550
CATV			1310	1550
Telephone			1310	1550
DWDM			1480-	1650
Fiber amplifiers			1480-	1650
WDM	850	1300	850	1550
Testing				1625
FTTH/PON			1310	1490 1550

 1- A Light Overview 14

14

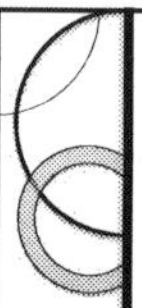

Light Properties (2-2)

- Wavelength
- Spectral width
- Speed
- Power
- Volume
- Pulse Width
- Polarization (Advanced Concept)
- Phase (Advanced Concept)
- Wavelength non-interference

 1- A Light Overview 15

15

Spectral Width (2-2)

- When we use the term 'wavelength', we imply, incorrectly, that light has a single wavelength
- In all communications systems, the opposite state is true
  - Light in the fiber includes a range of wavelengths centered around a 'central', or 'peak', wavelength
- This range is the 'spectral width' of the light

 1- A Light Overview 16

16

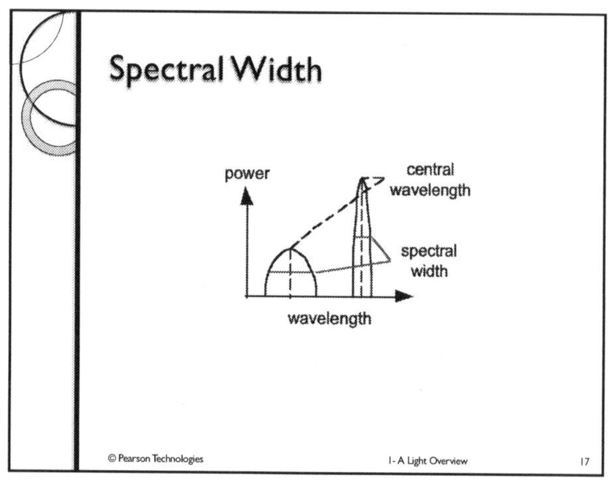

17

Significance

- Each wavelength within the spectral width travels at a different speed
  - Different speeds result in dispersion
- Spectral width is a specification for optoelectronics
  - More on this later (see Dispersion)

 1- A Light Overview 18

18

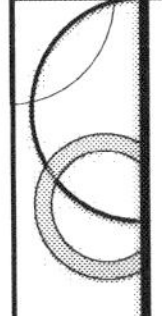

Light Properties (2-2)

- Wavelength
- Spectral width
- Speed
- Power
- Volume
- Pulse Width
- Polarization (Advanced Concept)
- Phase (Advanced Concept)
- Wavelength non-interference

19

Speed Of Light

- We all learned about the speed of light in school
  - The speed of light, c, is the speed at which light travels in a vacuum
- c= 2.994×10^8 m/sec (185,991 miles/second)
- When light travels in any material, its speed drops

20

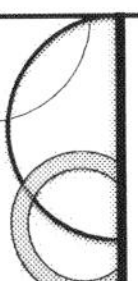

High Technologies..

- Use somewhat obscure terms for simple concepts
- Being a high technology, fiber optics will 'obscure' the term 'speed of light'
- For speed of light, we use
  - Refractive index (RI) or
  - η

21

Refractive Index (RI)

- Definition
  - RI= (speed of light in vacuum/speed of light in a material)
- The RI of optical fibers ranges from approximately 1.46 to 1.52 (Appendix 1)
- Significance: the RI is used to calibrate OTDR for accurate length and attenuation rate measurements (See OTDR Testing)

22

Typical RI Values

Product		850 nm	1300 nm	1310 nm	1550 nm
Draka MaxCap OM-2	50/125	1.482	1.477		
Draka MaxCap OM-3	50/125	1.482	1.477		
Draka MaxCap OM-4	50/125	1.482	1.477		
Laserwave 550-300	50/125	1.483	1.479		
LaserWave G+	50/125	1.483	1.479		
InfiniCor 50	50/125	1.481	1.476		
InfiniCor 62.5	62.5/125	1.496	1.491		
SMF-28e+				1.4676	1.4682
Draka singlemode G.652				1.4670	1.4680
OFS AllWave ZWP				1.4670	1.4680

23

Key Fact

- For a given core diameter or fiber type and wavelength, the RI will be approximately the same for fiber from all manufacturers
- Use of generic value will produce usable measurements

24

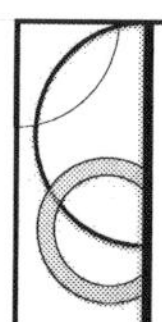

Light Properties (2-2)

- Wavelength
- Spectral width
- Speed
- Optical Power
- Volume
- Pulse Width
- Polarization (Advanced Concept)
- Phase (Advanced Concept)
- Wavelength non-interference

© Pearson Technologies 1- A Light Overview 25

25

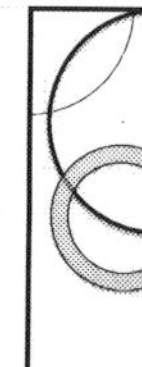

Optical Power

- We can measure
  - Power
  - Power loss
- If we measure power level, we measure absolute power
- If we measure power loss, we measure relative power

© Pearson Technologies 1- A Light Overview 26

26

Absolute Power Units

- Milliwatts
- dBm
  - A power level relative to one milliwatt
- Definition:
  - dBm= 10 log (power level/one milliwatt)
- Examples
  - 0 dBm= 1 milliwatt
  - 10 dBm= +10 milliwatts
  - 20 dBm= +100 milliwatts
  - -10 dBm= 0.1 milliwatts

© Pearson Technologies 1- A Light Overview 27

27

dBm Uses

- dBm is used as measure of power
  - Launched into a fiber by a transmitter
  - Delivered to a receiver
  - Required for the receiver to function properly

© Pearson Technologies 1- A Light Overview 28

28

Relative Power Units

- Relative power is measured in units of dB
- Definition
  - dB= 10 log (power level/arbitrary power level)
- In testing, the 'arbitrary' power is a
  - 'Reference' or 'Input' power level
- Used to indicate loss of power in
  - Fiber
  - Splice
  - Connector
  - Cable plant
  - Defects

© Pearson Technologies 1- A Light Overview 29

29

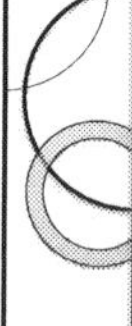

dB Examples

- We use the term dB to indicate
  - Power loss of a component in a fiber link
    - Connectors specified ≤ 0.75 dB/pair
    - Fiber attenuation rate specified ≤ 0.3-3.0 dB/km
  - Total loss in a fiber link (the loss budget)
  - Maximum loss that can occur between a properly functioning transmitter and receiver (the power budget)
    - E.g., transmitter-receiver maximum loss specified ≤ 12 dB

© Pearson Technologies 1- A Light Overview 30

30

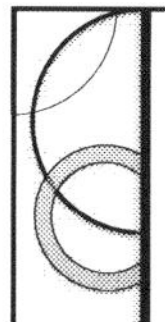

dB Examples

- dB=10 log(power factor)
- 10 dB=10 log(10)
- 20 dB=10 log(100)
- -10 dB=10 log(.10)
- -20 dB=10 log(0.01)

© Pearson Technologies 1- A Light Overview 31

31

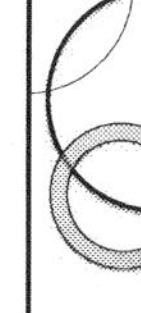

The Importance Of Power

- Power level drops as the light travels through link
- Power level at the input end of the link must be high enough that the receiver can convert the optical pulse to the same digital pulse at the transmitter, because…
  - As we all know, all electronic devices require a minimum power level to function properly

© Pearson Technologies 1- A Light Overview 32

32

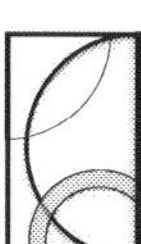

Light Properties (2-2)

- Wavelength
- Spectral width
- Speed
- Power
- Volume
- Pulse Width
- Polarization (Advanced Concept)
- Phase (Advanced Concept)
- Wavelength non-interference

© Pearson Technologies 1- A Light Overview 33

33

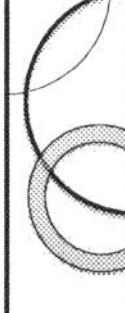

Volume

- In singlemode transmission, waves carry energy in the form of a field
- Energy fields exist within a volume (See: mode field diameter, MFD)

© Pearson Technologies 1- A Light Overview 34

34

Light Properties (2-3)

- Wavelength
- Spectral width
- Speed
- Power
- Volume
- Pulse Width
- Polarization (Advanced Concept)
- Phase (Advanced Concept)
- Wavelength non-interference

© Pearson Technologies 1- A Light Overview 35

35

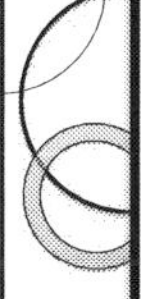

Pulse Width

- Fiber optics has two types of 'pulse width'
- Each transmission pulse has width
  - Width results from time to turn on and off a transmitter
  - At the receiver, this width must be less than that of the time interval allowed by the data rate
  - Gigabit example: at receiver, pulse width must be less than one billionth of a second
- In OTDR testing, pulse width determines length of fiber that can be tested (See OTDR Testing)

© Pearson Technologies 1- A Light Overview 36

36

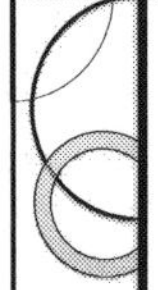

Light Properties (2-3)

- Wavelength
- Spectral width
- Speed
- Power
- Volume
- Pulse Width
- Polarization
- Phase (Advanced Concepts)
- Wavelength non-interference

37

Wavelength Non-Interference

- Different wavelengths do not interfere (corrupt)
  - They can be coupled into the fiber
  - They can be decoupled and separated
- Transmission of multiple wavelengths is known as wavelength division multiplexing (WDM)
  - WDM: 2 wavelengths, 1310nm and1550nm
  - CWDM, coarse wavelength division multiplexing: up to 8 wavelengths
  - DWDM, dense wavelength division multiplexing: up to 200 wavelengths
- DWDM wavelengths, centered around 1550nm, can be simultaneously amplified in an optical amplifier
  - EDFA, erbium doped fiber amplifier, was first such amplifier
  - Raman amplifier is second type

38

The Language Of Light

- Concepts
- Properties
- Behavior
- Numbers
- Significance of the different types of fiber (Chapter 3)

39

Light Behavior(2-3)

- Reflection
- Refraction
- Dispersion
- Attenuation
- Skew (Advanced Concept)

40

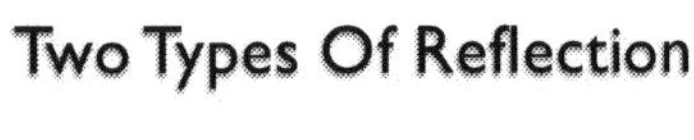

Two Types Of Reflection

- We all recognize reflections
- For example, we see the sky reflected in a lake, which is a total reflection
- We see ourselves dimly reflected when we look through a closed window, which is a partial reflection
- Both types of reflections explain the behavior of light in optical fibers and their connections

41

Total Reflection

- The sky is reflected in a lake because there is a change in the speed of light/RI at the air-water boundary
- This change results in reflection as long as the angle of reflection is proper
- In the terms of physics, we see the sky as long as the rays of light strike the water within a critical angle
- If we look closer and closer to our feet, at some increased angle, which we call the 'critical angle', we stop seeing the sky and see into the water

42

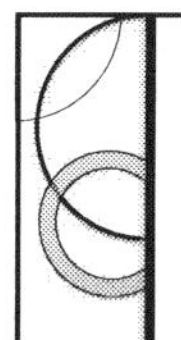

The Critical Angle

- The maximum angle to the surface of the water at which we can see the sky is the critical angle
- Multimode fibers confine light in core because of a critical angle
- Critical angle is a characteristic and specification of multimode fibers
- The cone created by the critical angle results in increased power loss in connectors when they are not in contact due to dirt on connector end (ferrule) (See Connectors)

© Pearson Technologies 1- A Light Overview 43

43

Critical Angle Reflection

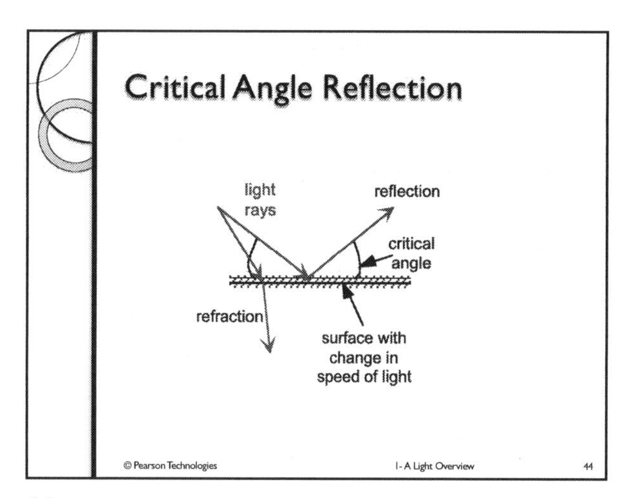

© Pearson Technologies 1- A Light Overview 44

44

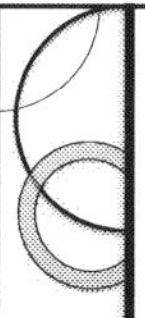

Numerical Aperture (Advanced Concept)

- High technologies use somewhat obscure terms for simple concepts
- For the term 'critical angle', we use
  - Numerical aperture or NA
- The NA, a measure of the critical angle, is defined
  - NA= sine (critical angle)
- Since the sine of an angle, in degrees, is a dimensionless number
  - NA is dimensionless
  - One of two dimensionless numbers in fiber optics

© Pearson Technologies 1- A Light Overview 45

45

Common NAs (Advanced Concept)

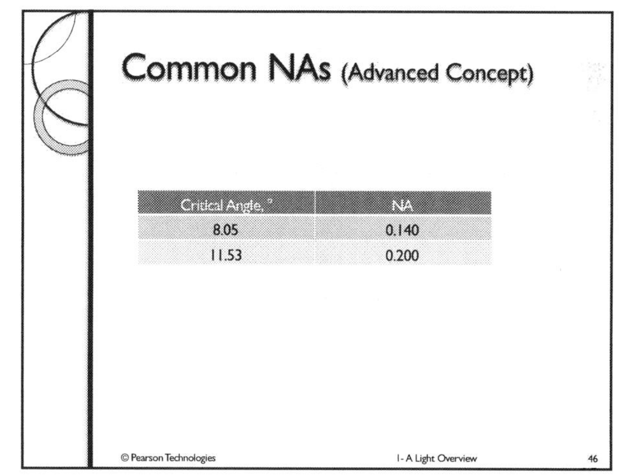

Critical Angle, °	NA
8.05	0.140
11.53	0.200

© Pearson Technologies 1- A Light Overview 46

46

Partial Reflection

- Our reflection in the surface of a closed window is a Fresnel reflection
- A Fresnel reflection is a partial reflection
- A Fresnel reflection occurs any time light moves from one material, with one RI, to another material, with a different RI
- Connectors and splices can create locations at which the RI can change due to imperfect surface smoothness
  - Resulting in reflections (See OTDR Testing)
- Significance: partial reflections can result in signal transmission errors

© Pearson Technologies 1- A Light Overview 47

47

Fresnel Reflections In Connectors

ferrule ferrule

surface roughness

© Pearson Technologies 1- A Light Overview 48

48

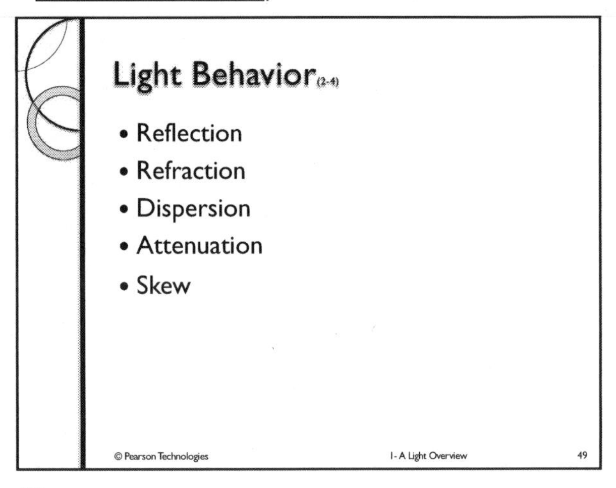

49

Refraction (2-4)

- Refraction, or bending, of light is common
- A pencil in a glass of water appears to be bent
- This bending occurs whenever light moves from a material, with one RI, to a material with a different RI
- The rule: as light moves from a higher RI to a lower RI material, it refracts towards the higher RI material
- Significance: bending explains the reduced dispersion that occurs in graded index multimode (GI) fibers (See Fiber)

© Pearson Technologies 1- A Light Overview 50

50

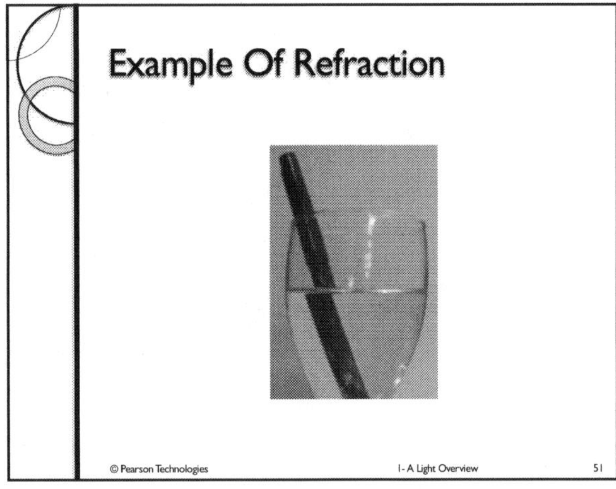

51

Light Behavior(2-4)

- Reflection
- Refraction
- Dispersion
- Attenuation
- Skew

© Pearson Technologies 1- A Light Overview 52

52

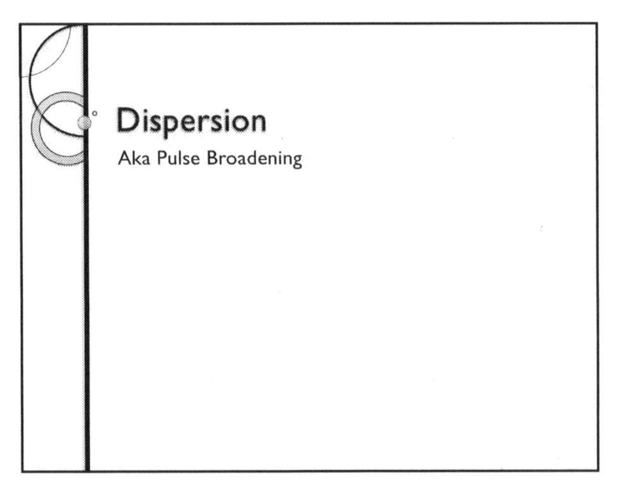

53

Key Concept

- Optical energy enters fiber at one time
- This energy exits the fiber over a range of times
- This is dispersion

© Pearson Technologies 1- A Light Overview 54

54

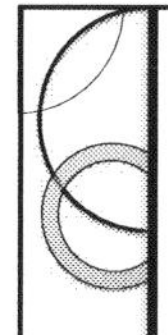

Key Fact

- Each optical pulse in fiber gets broader in time (it disperses) as it travels further along the fiber, resulting in a limitation on transmission distance

55

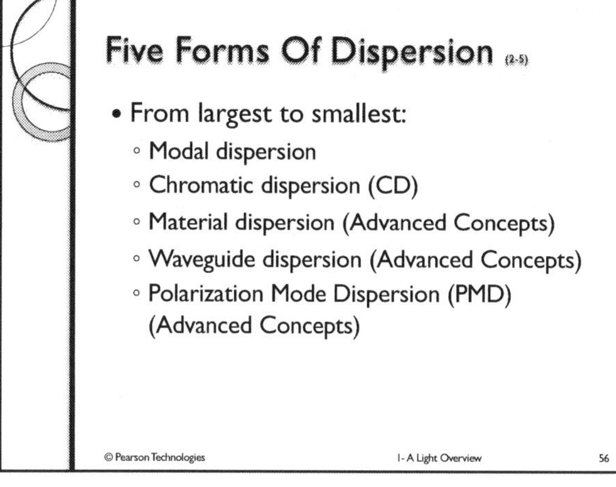

56

Modal Dispersion

- Definition: a path is a 'mode'
- Modal dispersion is path dispersion
  - All light rays in a pulse enter the fiber at the same time
  - Rays can travel multiple paths
    - This dispersion occurs in multimode (multi-path) fibers, but not in singlemode (single path) fibers
- All rays can travel the same path length
    - Single path means single mode

57

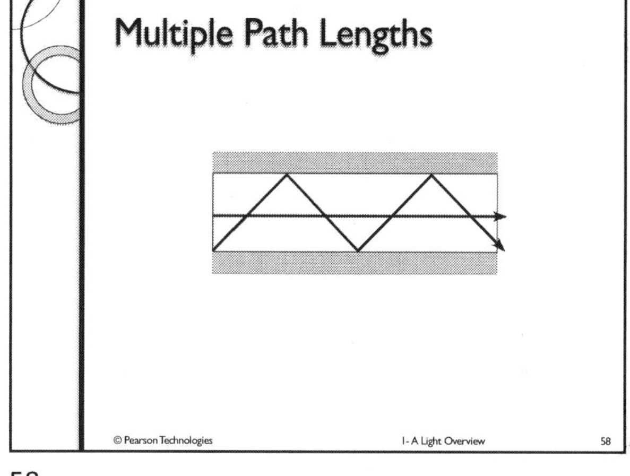

58

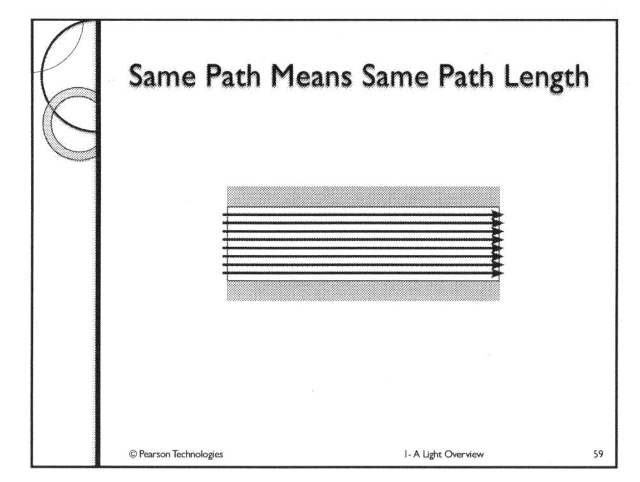

59

Same Path Length

- Results in 'single path' transmission
- We call this singlemode transmission
- Caution: same path length does not mean that all rays arrive at end of fiber at same time

60

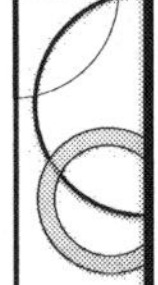

Chromatic Dispersion

- Even if all rays travel same path, all rays may not arrive at same time
- Spectral width of transmitter creates multiple wavelengths, each traveling at a slightly different speed (See Spectral Width)
- Spectral width results in second largest cause of dispersion
  - Known as chromatic dispersion
- Chromatic dispersion occurs in both multimode and singlemode fibers

© Pearson Technologies 1- A Light Overview 61

61

Dispersion And Accuracy (2-6)

- Let us examine digital transmission
- In all digital transmission systems, we have a thresh hold power level
- At the receiver, power levels below the threshold result in a signal 'zero'
- At the receiver, power levels above the threshold, result in a signal 'one' (RZ)
- This threshold may be at zero power or at some level above zero (NRZ)

© Pearson Technologies 1- A Light Overview 62

62

Threshold

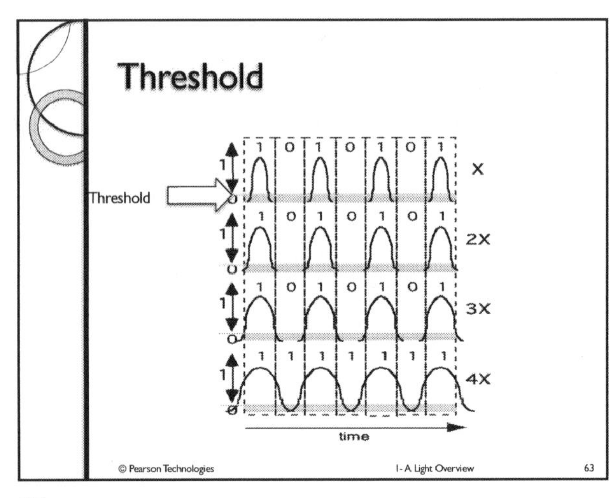

© Pearson Technologies 1- A Light Overview 63

63

Dispersion Means Delayed Energy Arrival

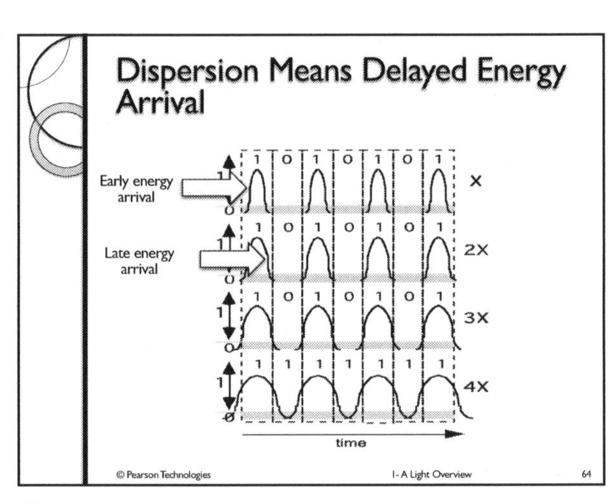

© Pearson Technologies 1- A Light Overview 64

64

Excessive Pulse Dispersion Results in Signal Inaccuracy

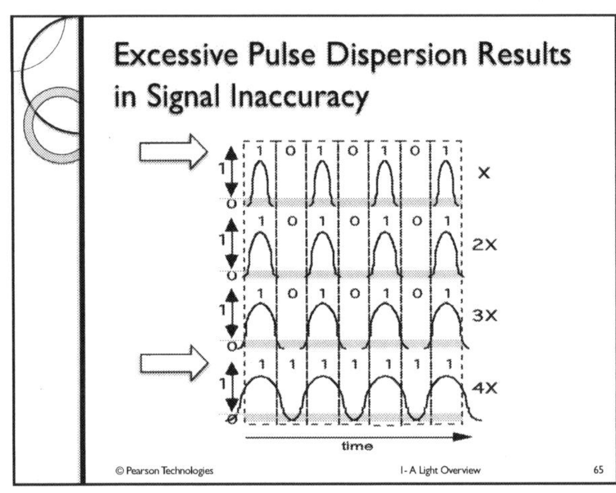

© Pearson Technologies 1- A Light Overview 65

65

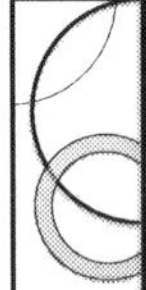

Pulse Width Increases With Distance

- As the fiber transmission distance increases, so does the pulse width at the receiver
- As pulse dispersion increases, energy from one pulse will arrive at the receiver in a time interval of an adjacent pulse (2X)
- If the pulse dispersion is excessive, enough energy from one pulse will arrive at the receiver in a time interval of an adjacent pulse
- With excessive pulse dispersion, a zero at the transmitter becomes a one at the receiver (4X)

© Pearson Technologies 1- A Light Overview 66

66

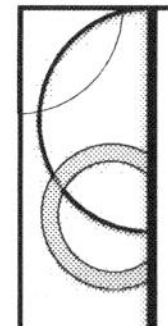

Installer Ignores Dispersion

- Despite the importance of pulse dispersion, the installer need not be concerned
- Dispersion is controlled by the size of the core, by the manner in which light enters the core, by the spectral width, and by atomic level composition variations in the core
- The installer can affect none of these characteristics to cause increased dispersion
- However, installer needs to know that dispersion limits transmission distance

67

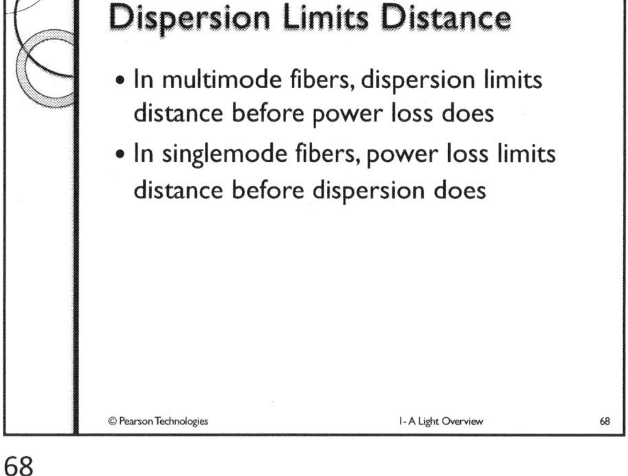

68

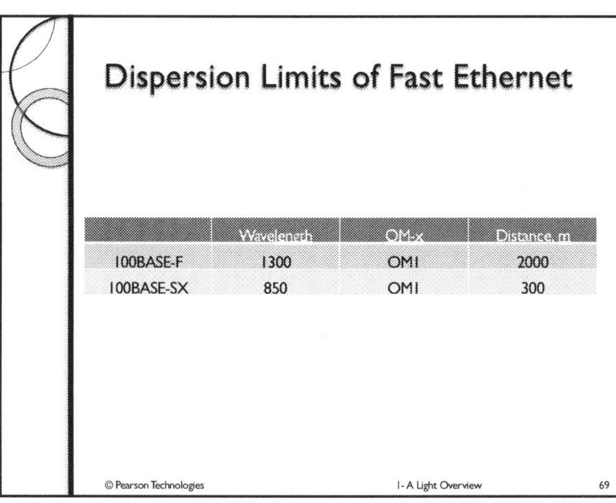

	Wavelength	OM-x	Distance, m
100BASE-F	1300	OM1	2000
100BASE-SX	850	OM1	300

69

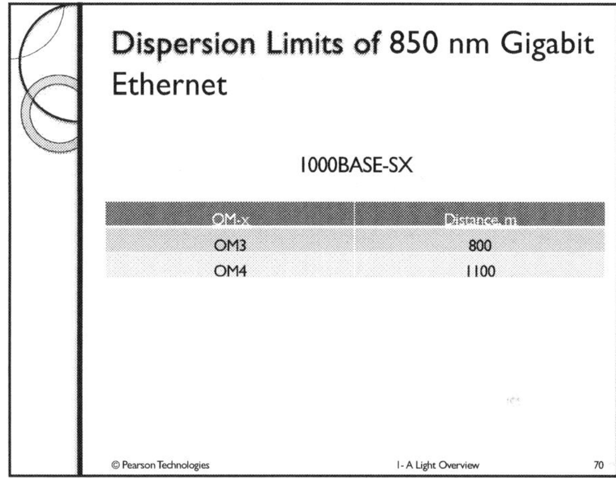

OM-x	Distance, m
OM3	800
OM4	1100

70

Limits of Gigabit Ethernet

1000BASE-LX

Wavelength	OM-x	Distance, m
1300	OM3	550
1300	OM4	550
1310	SM	5000
1310	SM	10000
1550	SM	~70000

Multimode distance is limited by dispersion; singlemode, by power loss.

71

Multimode to Singlemode

- Note significant distance increase possible with change from multimode to singlemode fiber
- Distance increase results from elimination of modal dispersion, the largest form of dispersion

72

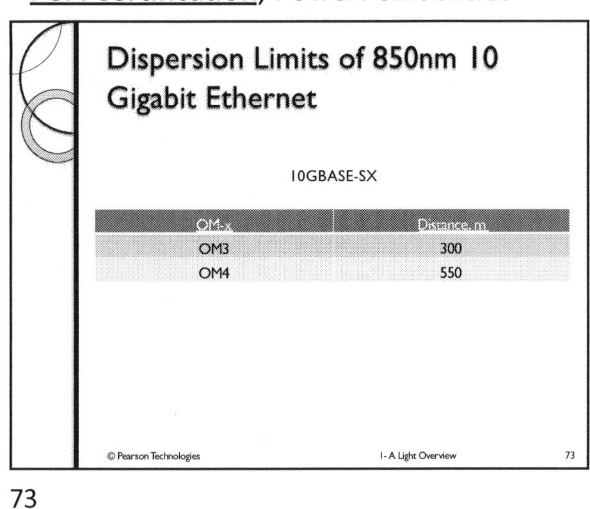

Dispersion Limits of 850nm 10 Gigabit Ethernet

10GBASE-SX

OM-x	Distance, m
OM3	300
OM4	550

© Pearson Technologies 1- A Light Overview 73

73

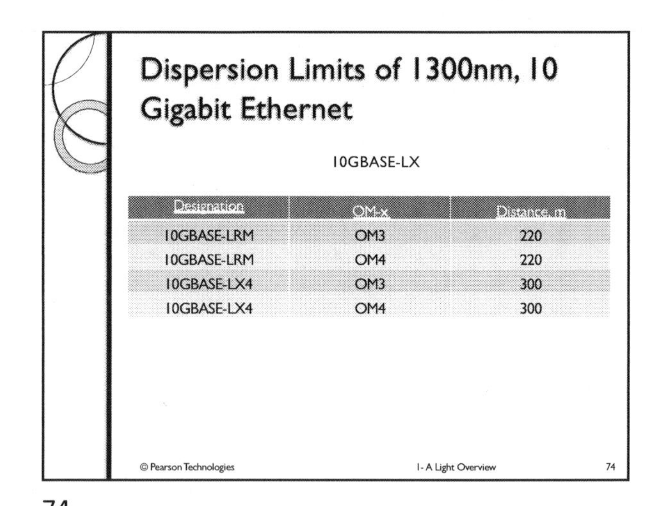

Dispersion Limits of 1300nm, 10 Gigabit Ethernet

10GBASE-LX

Designation	OM-x	Distance, m
10GBASE-LRM	OM3	220
10GBASE-LRM	OM4	220
10GBASE-LX4	OM3	300
10GBASE-LX4	OM4	300

© Pearson Technologies 1- A Light Overview 74

74

Dispersion Limits of 850nm, 40 and 100 Gigabit Ethernet

40GBASE-SX, 100GBASE-SX

Designation	OM-x	Distance
40GBASE-SR4	OM3	100
40GBASE-SR4	OM4	125
100GBASE-SR10	OM3	100
100GBASE-SR10	OM4	125

© Pearson Technologies 1- A Light Overview 75

75

Light Behavior

- Reflection
- Refraction
- Dispersion
- Attenuation
- Skew

© Pearson Technologies 1- A Light Overview 76

76

Attenuation

- Is power loss
- All network components have power loss
- In fiber, we refer to power loss as attenuation with a specification of attenuation rate (See Fiber)
- In connectors and splices, we refer to power loss as 'loss'

© Pearson Technologies 1- A Light Overview 77

77

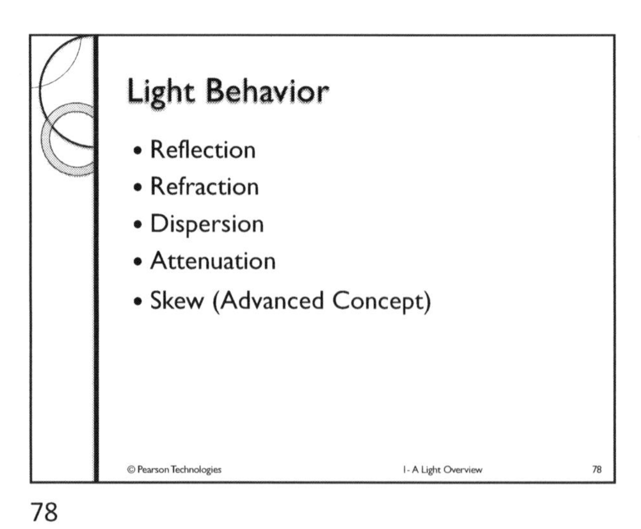

78

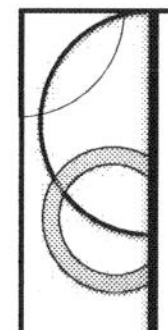

Skew (Advanced Concept)

- The RI is a nominal value, not an exact value
- Two optical signals on separate fibers will have slightly different RIs and travel times
- The difference in travel times between fibers is optical 'skew'
- Skew is a specification for multimode cables used for 40G and 100G Ethernet transmission

© Pearson Technologies 1- A Light Overview 79

79

40GBASE-SX, 100GBASE-SX

- 'SX' designates multimode transmission at 850 nm
- The 40Gbps electrical signal is de-multiplexed into 4 parallel, 10 Gbps signals
- The 100 Gbps electrical signal is demultiplexed into 10 parallel, 10 Gbps signals
- Each 10 Gbps signal is transmitted on a separate fiber

© Pearson Technologies 1- A Light Overview 80

80

100GBASE-SX

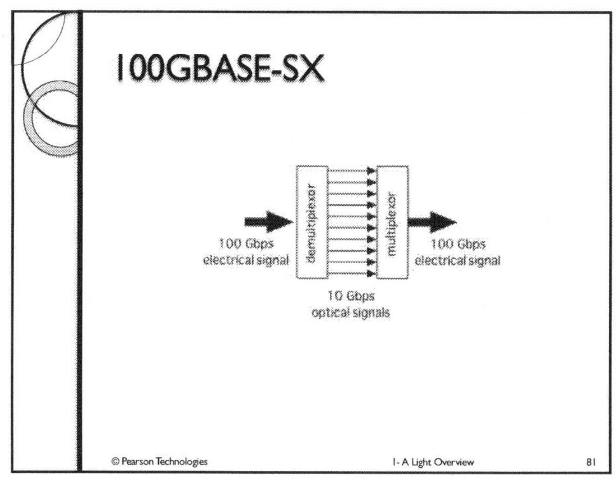

© Pearson Technologies 1- A Light Overview 81

81

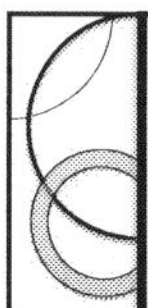

Summary

- Light has characteristics, with performance values and units of measure

© Pearson Technologies 1- A Light Overview 82

82

Characteristics, Terms, Units

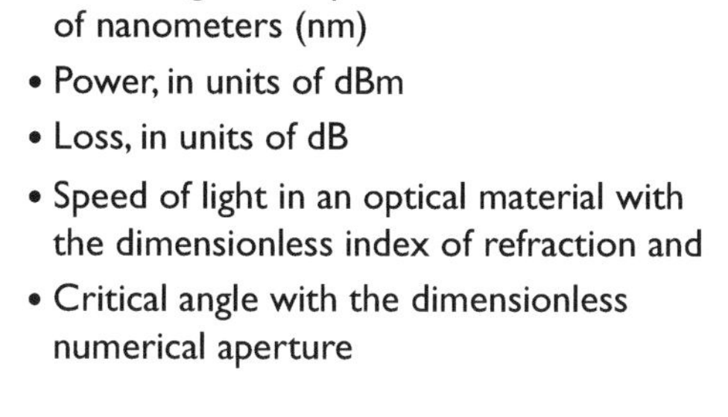

- Wavelength and spectral width with units of nanometers (nm)
- Power, in units of dBm
- Loss, in units of dB
- Speed of light in an optical material with the dimensionless index of refraction and
- Critical angle with the dimensionless numerical aperture

© Pearson Technologies 1- A Light Overview 83

83

End Of Light

Questions?

Comments?

Observations?

84

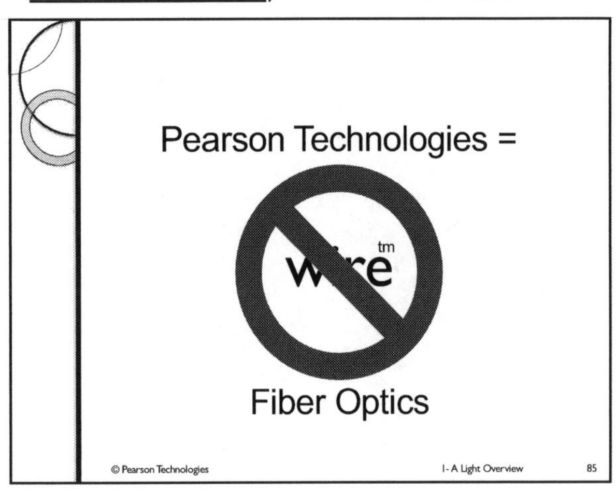
Pearson Technologies =
wire
tm
Fiber Optics
© Pearson Technologies
1- A Light Overview
85

85

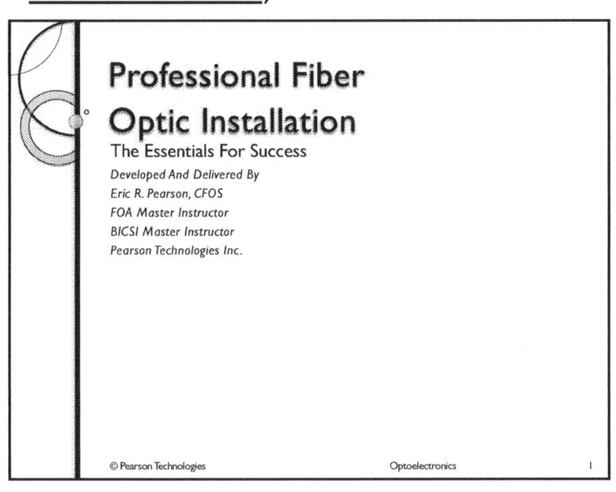

1

Optoelectronics

Without these, the links are wasted!

© Pearson Technologies Optoelectronics 2

2

Objective

- In this chapter, you learn optoelectronic
  - Definition
  - Function
  - Language
  - Types
  - Numbers
  - Definition
  - Active devices
  - Performance

© Pearson Technologies Optoelectronics 3

3

Definition (8-1)

- Optoelectronics are the devices on the ends of the link
- Optoelectronics includes all devices that convert signals
  - Media converters, SFP modules, transponders
- The term 'optoelectronics' describes both the transmitter and receiver, since both function with both electrical and optical signals

© Pearson Technologies Optoelectronics 4

4

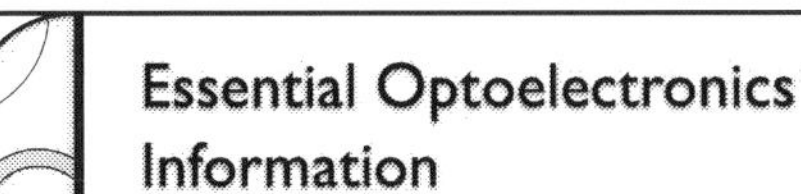

- Definition
- Function
- Active devices
- Performance
- Configurations
- OPBR calculation
- Excess power calculation
- OPBR statement

© Pearson Technologies Optoelectronics 5

5

Function= Conversion

- The transmitter converts an electrical signal to an optical signal; the receiver performs the reverse conversion
- Conversion is done in the 'active device'
- The key design concern is accuracy of signal conversion

© Pearson Technologies Optoelectronics 6

6

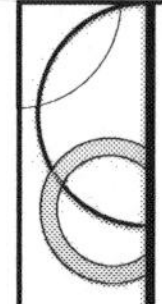

Accuracy Achieved When..

- The power level at the receiver is proper
  - High enough
  - Not excessive
- The pulse dispersion is sufficiently low
- In coherent transmission links, the optical signal to noise ratio (OSNR) is above a minimum value (Advanced Concept)

© Pearson Technologies Optoelectronics 7

7

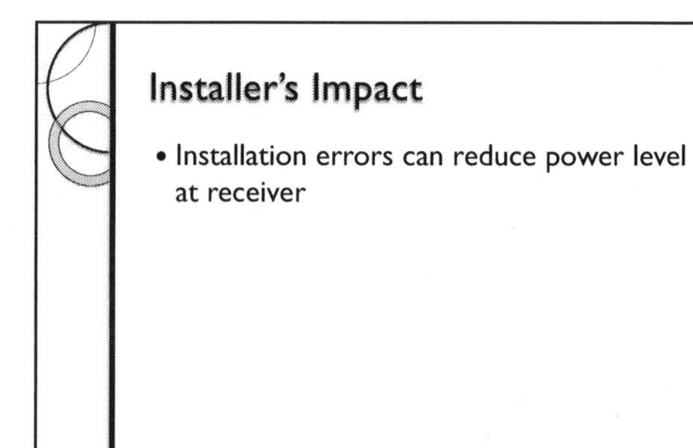

Installer's Impact

- Installation errors can reduce power level at receiver

© Pearson Technologies Optoelectronics 8

8

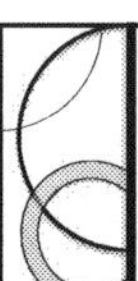

Installer Can't Degrade Dispersion

- Dispersion controlled by factors that installer cannot change

© Pearson Technologies Optoelectronics 9

9

Transceivers

- Transmit and receive simultaneously
- Have many designations
  - Media Converters
  - GBICs
  - SFP (1 Gbps)
  - SFP+ (10 Gbps)
  - XFP
  - XENPAK
  - QSFP
  - QSFP+
  - CFP

© Pearson Technologies Optoelectronics 10

10

Media Converters

- Used to convert copper networks to fiber
- Extend the transmission distance of copper networks
- Eliminate RF and EM interference

© Pearson Technologies Optoelectronics 11

11

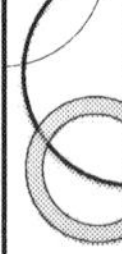

Designations

- -SX, 850 nm, multimode
- -LX, 1310 nm, 10 km
- Extended distance designations below can overload receiver
  - -EX, 1310 nm, 40 km
  - -ZX, 1550 nm 80 km
  - -EZX, 1550 nm, 120 km
  - -BX, 1490/1310 nm, 10 km
- -SFSW, bidirectional on one fiber

© Pearson Technologies Optoelectronics 12

12

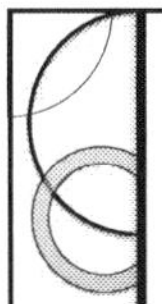

Transponders

- Convert one wavelength to another wavelength
- Used in CWDM and DWDM networks
- Used after transmitter and before start of transmission path

 Optoelectronics 13

13

Essential Optoelectronics Information

- Definition
- Function
- Active devices
- Performance
- Configurations
- OPBR calculation
- Excess power calculation
- OPBR statement

 Optoelectronics 14

14

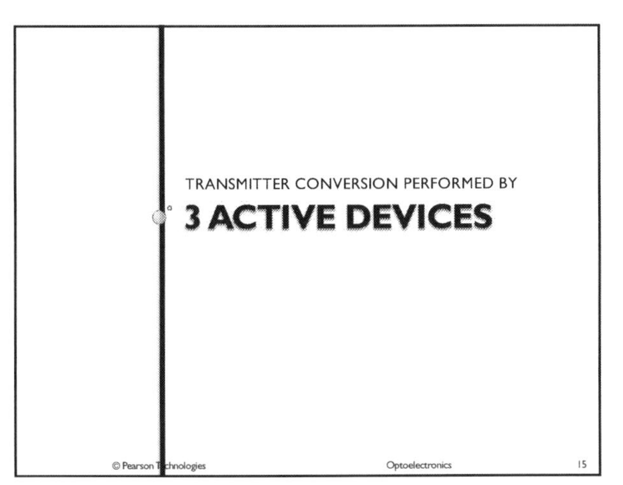

15

Three Transmitter Active Devices

- Light emitting diode (LED)
  - Multimode fiber at 850nm
- Laser diode (LD)
  - Singlemode fiber
  - Some multimode, 1300nm links (recent addition)
- Vertical cavity, surface emitting laser (VCSEL)
  - ~850nm at 1 Gbps and 10 Gbps

 Optoelectronics 16

16

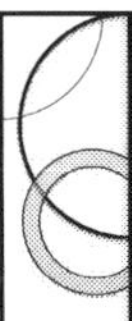

LED Characteristics

- Large spot size (150-250 μm)
- Circular spot shape
- Relatively low bit rate/ bandwidth capability (< 200 Mbps)
- Relatively low launch power (1-10 μW)
  - Lowest output power of 3 transmitter active devices
- Relatively low cost

 Optoelectronics 17

17

Use

- Used in multimode transmission at 850 nm and 1300 nm
- Low bit rate ~ < 200 Mbps
  - 100 Mbps common with LED
- Short distance ≤ 2 km
- Multimode

 Optoelectronics 18

18

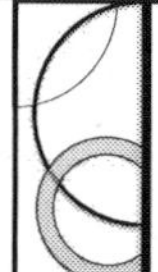

LD

- Small spot size (2μ x10μ)
- Relatively high bit rate/ bandwidth capability (~25 Gbps)
- Highest output power (>1 mW) of 3 transmitter active devices
- Singlemode transmission from 1310nm to 1550nm
- Fabry Perot (F-P) lasers operate at 1310nm
- Distributed feedback (DFB) lasers operate at 1310nm to 1550nm

© Pearson Technologies Optoelectronics 19

19

LD Dimensions

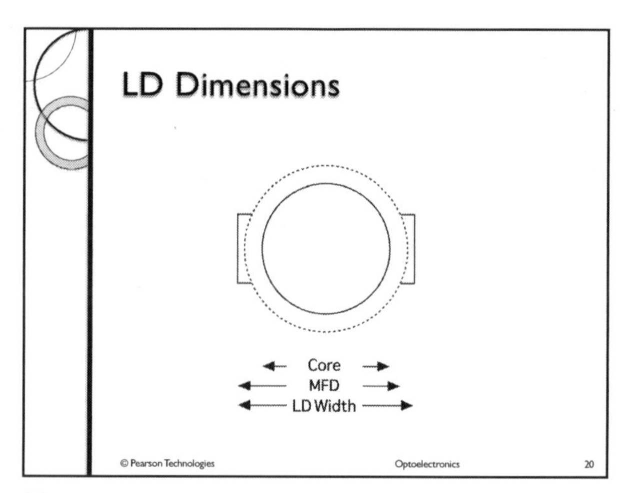

© Pearson Technologies Optoelectronics 20

20

Use

- High bit rate
- Long distance
- Singlemode links
  - 1310 nm
  - 1550 nm
  - 1490 nm
  - 1310-1550 nm

© Pearson Technologies Optoelectronics 21

21

LD Power Level Misleading

- DWDM and CWDM have multiple wavelengths, resulting in high total power level
- Specific case of FTTH
  - 1550 nm wavelength can be amplified to high level, ~20 dBm
  - Safety concern: eye damage possible

© Pearson Technologies Optoelectronics 22

22

VCSEL

- Medium spot size, 30μ
- Annular spot shape
- Small angle of divergence
- Relatively high bit rate
  - 1 and 10 Gbps
- Relatively high launch power
- Relatively low cost
- 850nm multimode VCSELs common

© Pearson Technologies Optoelectronics 23

23

Unique Characteristic

- Restricted modal launch
  - Annular spot shape
  - Launches very little power in center of fiber
  - Launch reduces modal dispersion
  - Results in increased bandwidth and distance capability, more than that possible with LED
- Shape eliminates splitting of pulse

© Pearson Technologies Optoelectronics 24

24

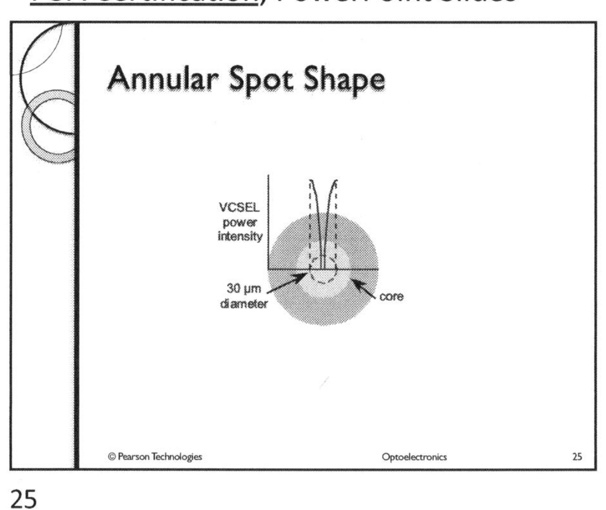

25

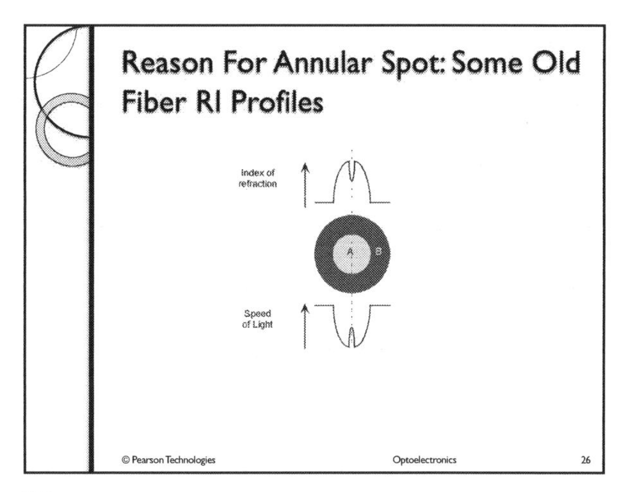

26

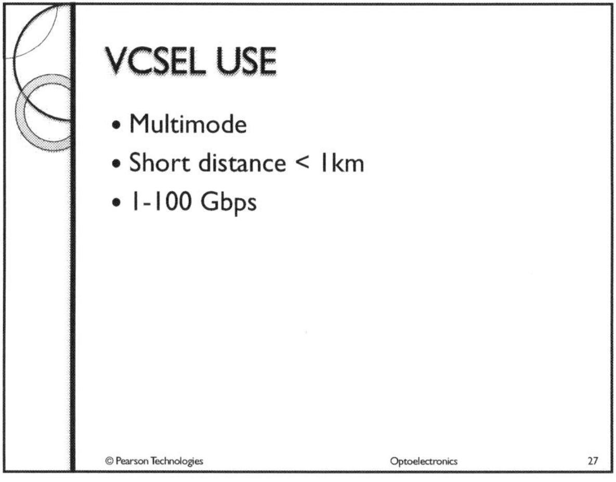

27

Two Active Devices Dominate

- Both are lasers
- VCSEL
  - On multimode fiber at 1 and 10 Gbps
  - In arrays for multimode transmission at 40 and 100 Gbps
- LD on singlemode fiber

© Pearson Technologies Optoelectronics 28

28

RECEIVER CONVERSION PERFORMED BY

2 ACTIVE DEVICES

© Pearson Technologies Optoelectronics 29

29

Two Receiver Active Devices

- Photodiodes
  - Used in data networks
  - Silicon detectors used for short wavelengths (660nm, 850nm)
  - InGaAs detectors used for long wavelengths (≥ 1300nm)

© Pearson Technologies Optoelectronics 30

30

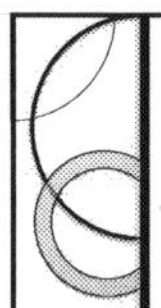

Fiber Amplifier

- Unique passive device as it functions as a receiver and transmitter
- The fiber amplifier amplifies the optical signal in the optical regime, without making the optical-electrical-optical conversions
- First type: erbium doped fiber amplifier (EDFA)
  - Operate at wavelengths in range of 1480-1650
- Second type: Raman amplifier

© Pearson Technologies Optoelectronics 31

31

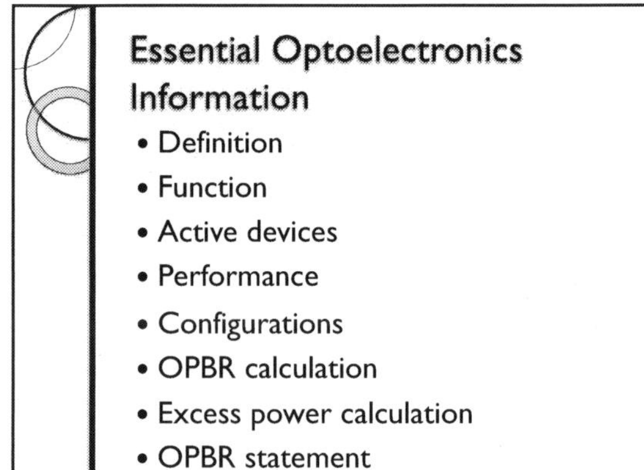

© Pearson Technologies Optoelectronics 32

32

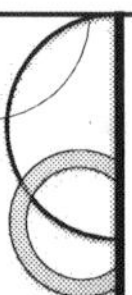

2 Or 3 Performance Requirements

- Wavelength
- Optical power budget (OPB)
- Minimum required loss (sometimes)

© Pearson Technologies Optoelectronics 33

33

Wavelength

- Required to perform testing: must test at wavelength of optoelectronics
  - To simulate operation

© Pearson Technologies Optoelectronics 34

34

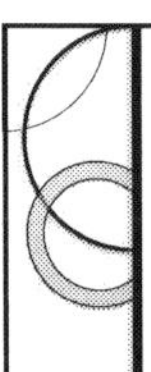

OPB (8-2)

- Is the maximum loss that can occur between a transmitter and receiver
- Value depends on standard
- The following tables include examples of various OPBs

© Pearson Technologies Optoelectronics 35

35

OPB For Fiber Standards

Standard	Core, µ	λ, nm	OPBA, dB
10BASE-FB	62.5	850	12
10BASE-FL	62.5	850	12
10BASE-FP	62.5	850	16-26
100BASE-F	62.5	1300	11
100BASE-SX	62.5	850	4.0
1000BASE-S	62.5	850	2.33
1000BASE-S	62.5	850	2.53
1000BASE-S	50	850	3.25
1000BASE-S	50	850	3.43
1000BASE-L	62.5	1300	2.32
1000BASE-L	50	1300	2.32

© Pearson Technologies Optoelectronics 36

36

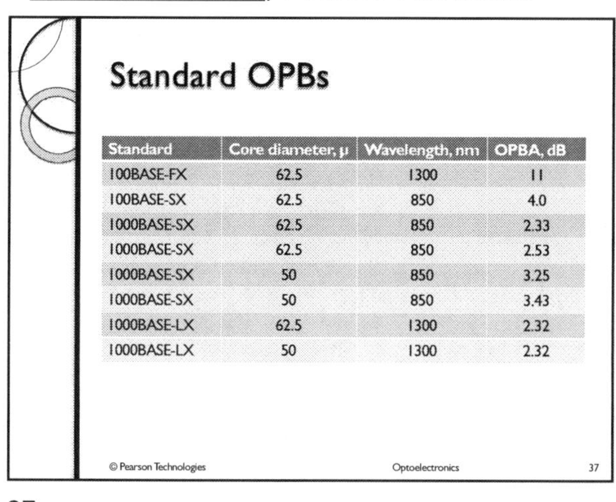

Standard OPBs

Standard	Core diameter, μ	Wavelength, nm	OPBA, dB
100BASE-FX	62.5	1300	11
100BASE-SX	62.5	850	4.0
1000BASE-SX	62.5	850	2.33
1000BASE-SX	62.5	850	2.53
1000BASE-SX	50	850	3.25
1000BASE-SX	50	850	3.43
1000BASE-LX	62.5	1300	2.32
1000BASE-LX	50	1300	2.32

© Pearson Technologies Optoelectronics 37

37

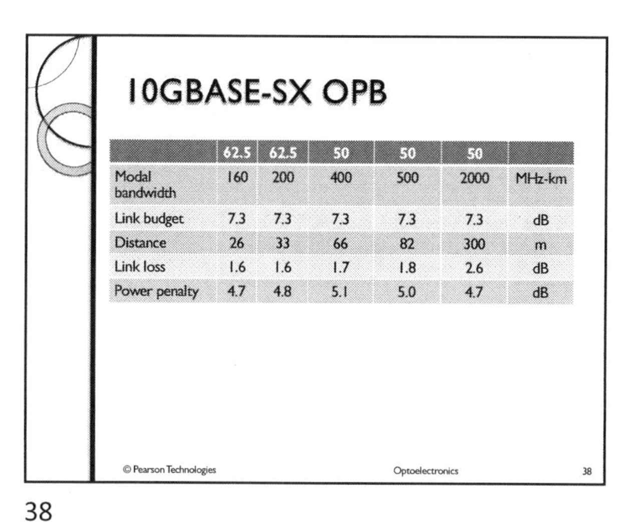

10GBASE-SX OPB

	62.5	62.5	50	50	50	
Modal bandwidth	160	200	400	500	2000	MHz-km
Link budget	7.3	7.3	7.3	7.3	7.3	dB
Distance	26	33	66	82	300	m
Link loss	1.6	1.6	1.7	1.8	2.6	dB
Power penalty	4.7	4.8	5.1	5.0	4.7	dB

© Pearson Technologies Optoelectronics 38

38

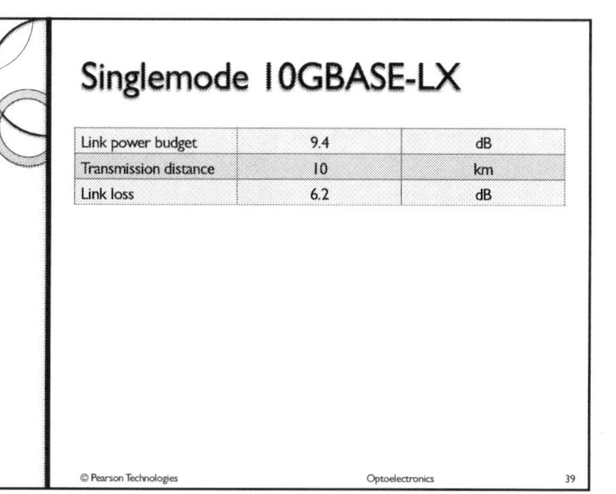

Singlemode 10GBASE-LX

Link power budget	9.4	dB
Transmission distance	10	km
Link loss	6.2	dB

© Pearson Technologies Optoelectronics 39

39

Required Minimum Loss

- Required minimum loss is the minimum loss that must occur between a transmitter and receiver to avoid overloading
- For most standards, RML=0 dB
- For extended distance, singlemode optoelectronics, RML> 0 dB
  - Extended distance optoelectronics can overload receiver on short links

© Pearson Technologies Optoelectronics 40

40

Direct Vs. Coherent Detection

- Advanced Concept
- Power loss is main concern of installer
- This is true up to 10 Gbps
- Direct detection is method of recovery of electrical signal in form of '1's and '0's
- Above 10 Gbps, OSNR becomes important
- OSNR important with optoelectronics that function with coherent detection and have optical amplifiers in link
- Coherent detection enables transmission of up to 8 electrical signal bits with one optical bit/pulse

© Pearson Technologies Optoelectronics 41

41

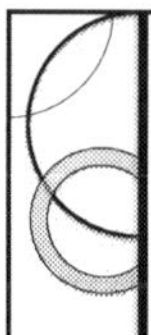

Installers Concerns

- Wavelength or wavelengths
- OPB
- With extended distance singlemode transmitters, required minimum loss (RML)
- Rule: keep transceivers and patch panel ports plugged when fibers not plugged in

© Pearson Technologies Optoelectronics 42

42

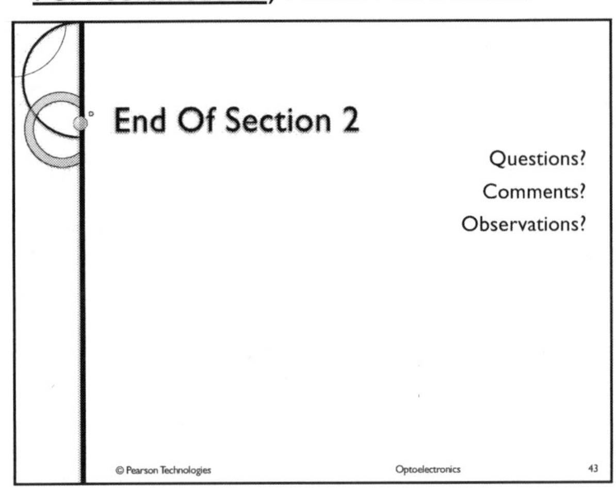
End Of Section 2
Questions?
Comments?
Observations?
© Pearson Technologies
Optoelectronics
43

43

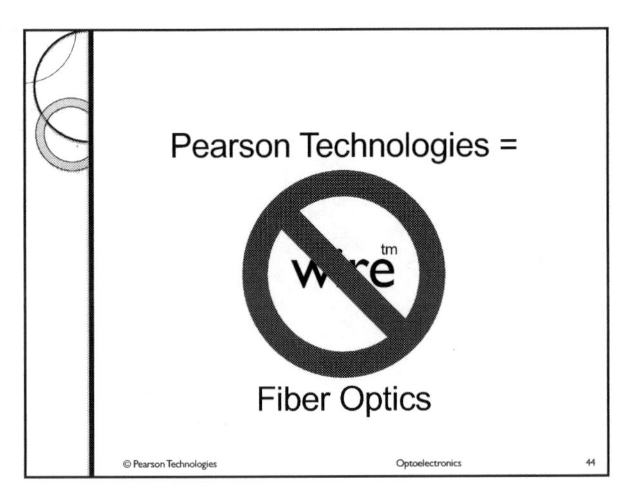
Pearson Technologies =
wire
tm
Fiber Optics
© Pearson Technologies
Optoelectronics
44

44

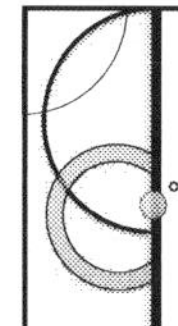

Professional Fiber Optic Installation

The Essentials For Success

Developed And Delivered By
Eric R. Pearson, CFOS
FOA Master Instructor
BICSI Master Instructor
Pearson Technologies Inc.

1

Fiber Optic Basics

A High-Level Overview

2

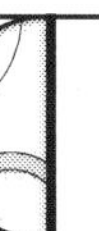

Fiber Basics

- Advantages
- Key Concepts
- Network Components
- Fiber Network Types
- Safety Concerns

3

Nine Advantages

- Nearly unlimited bandwidth
- Long transmission distance
- EMI/RFI immunity
- Lowest cost/bit
- Dielectric construction
- Small size
- Light weight
- Easy installation
- Security

4

Nearly Unlimited Bandwidth

- Singlemode fiber has (nearly) unlimited bandwidth, resulting in lowest cost/bit of any medium
  - Multimode fiber is limited to ~10 Gbps
- One fiber can transmit multiple singlemode wavelengths, each carrying
  - 25 Gbps/wavelength now
  - 40 Gbps/wavelength in near future
  - ≥ 200 Gbps/wavelength with coherent transmission
  - High capacity/bandwidth reduces the number of fibers and cables required

5

Coherent Transmission

- 8 electrical data bits/optical bit
- 25 Gbps/optical= 200 Gbps/electrical per wavelength
- 200 wavelengths in G.692
- 200 wavelengths @200 Gbps/wavelength
  - =40000 Gbps/fiber= 40 Tbps/fiber
  - Today
  - Tomorrow? 1 Quadrillion? 10^{15} 1,000,000,000,000,000 or one thousand, million, million bits

6

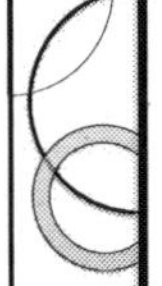

Metropolitan Networks Require High Bandwidth

- Such networks provide multiple services including
  - Surveillance cameras (CCTV)
  - Emergency service communications
  - Traffic monitoring and control
  - Educational systems and services
  - Telephone
  - LAN
  - Security
  - Leased communication services on dark fibers

7- Design 7

7

Long Transmission Distance 1

- Enabled by
  - Singlemode fiber
  - Optical amplification (up to 100 km spacing)
  - Dispersion compensation
- In outside plant (OSP), long distance transmission results in the lowest cost/bit of any transmission technology
- Champion Example
  - Alcatel-Lucent (7/12/13)
  - 7200 km= 4464 miles without return to electrical regime
  - Or approximately New York City to Rome Italy in optical regime
  - (https://www.cablinginstall.com/cable/article/16474217/alcatellucent-sets-new-singlefiber-optical-transmission-record-at-transoceanic-distance)

© Pearson Technologies 1- Basics 8

8

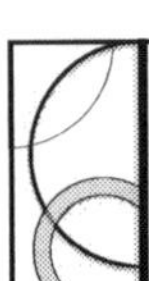

Long Transmission Distance 2

- Long transmission distance reduces network cost through reduction in
  - Number of repeaters
  - Power requirements
  - Maintenance cost

© Pearson Technologies 1- Basics 9

9

Long Transmission Distance 3

- In local area networks, long transmission distance reduces network cost through reduction or elimination of telecom closets and their support costs
  - (Cooling, UPS, stable power, security, real estate value, hardware)
- Centralized cabling architecture requires no closets

© Pearson Technologies 1- Basics 10

10

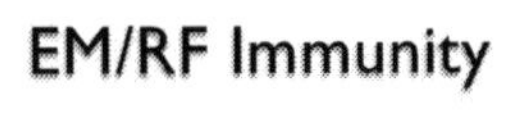

EM/RF Immunity

- Fiber is immune to interference from
  - Electromagnetic (EM) noise in environment
  - Radio frequency (RF) noise in environment
- Enables noise free transmission
- Enables long distance transmission
- Immunity is a major reason for use of fiber in industrial and process control networks

© Pearson Technologies 1- Basics 11

11

Lowest Cost/Bit Enabled by

- Multiple wavelengths
- Optical amplifiers
- Long distance without need for signal regeneration
- In fiber to the home (FTTH), prefabricated systems reduce installation time and cost
  - FTTH is one of fastest growing segments of fiber industry
- Reduction in 'support' costs in LANs
- Increased bandwidth with change of end optoelectronics
- Long life cycle/low life cycle cost
  - 30-year cable life not unusual

© Pearson Technologies 1- Basics 12

12

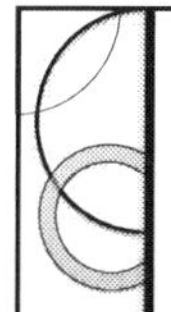

Dielectric Construction

- Eliminates cost for grounding and bonding
  - Elimination reduces both initial installation and maintenance costs
- Increases safety of system

13

Small Size 1

- 1, 1" diameter fiber optic cable can replace more than 37,305 3" diameter, 900 pair cables (Author's estimate)
- Increase the capacity of existing conduits
  - A key advantage in metropolitan networks helps avoidance of high cost of installing additional conduits

14

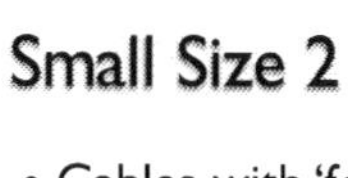

Small Size 2

- Cables with 'foldable' ribbons contain
  - 3456 fibers in diameter of 1.34"
  - 6912 fibers in diameter of 1.38"
  - Each fiber pair can support 28 Tbps
  - This cable capacity: 96,768 Tbps or 172,800,000,000,000 simultaneous telephone conversations for 7,674,000,000 people in the world (2019, World Bank)
  - Everyone on earth could talk to everyone else with a single, 1.38" diameter cable

15

Light Weight

- Light weight results in use by Army, Navy, and Air Force
- Army: soldiers can carry increased cable length to isolate front lines from soldiers monitoring front lines; separate radar dishes from soldiers
- Navy: reduce weight above waterline, increasing ship stability
- Air Force: reduce plane weight, increasing mission endurance

16

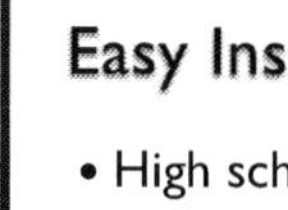

Easy Installation

- High school students can install cables and connectors
- Cables resistant to abuse
- No polish connectors provide consistent low loss
- Splicers provide automatic low loss, high strength splices

17

Security

- No radiation of signals
- Extremely difficult (nearly impossible) to tap signals without tapping being detected
  - Example: NSA 65-6
- No EMI, RFI, jamming possible
- Fiber can be used as security sensor

18

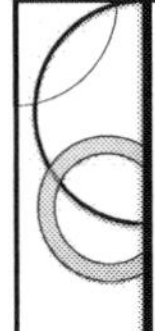

Advantages of Fiber

- Favor use in backbones
  - Telephone
  - CATV
  - Internet
  - Cell phones
  - Campus networks

© Pearson Technologies 1- Basics 19

19

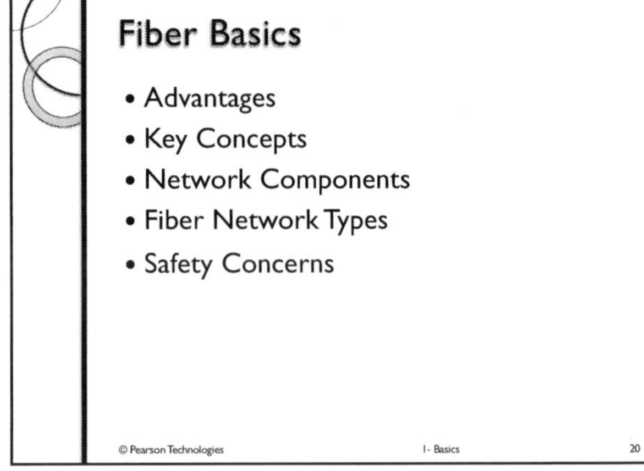

20

Key Concepts 1

- Communication is 'full duplex'
  - Duplex means simultaneous transmission and reception
- Consequence: 2 fibers required for most transmission systems
- Common problem: 'crossed fibers'
  - Solution: use visual fault locator (VFL) to verify polarity

© Pearson Technologies 1- Basics 21

21

Key Concepts 2

- Transmission can be
  - Digital
  - Analog
    - Initially, CATV systems were analog
    - Analog transmission distance limited ~10-20km
- Most fiber systems are digital
  - Digital transmission enables significantly increased distance, up to 120km (74.4 mi), without amplifier

© Pearson Technologies 1- Basics 22

22

Key Concepts 3

- Light travels in fiber
- Light has color, known as wavelength
- Wavelengths are infrared wavelengths, which are invisible to the eye
  - Human eye can see up to 780nm (https://en.wikipedia.org/wiki/Visible_spectrum)
  - Some people can see up to 850nm
  - Optical power in fiber is high enough to damage eyes
  - Consequence: eye protection required when viewing connectors

© Pearson Technologies 1- Basics 23

23

Key Concepts 4

- Most data networks are designed in compliance with the Building Wiring Standard, TIA/EIA-568
  - This standard defines copper and fiber cable systems
- This standard defines the 'hierarchical star' as the preferred topology

© Pearson Technologies 1- Basics 24

24

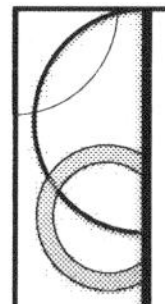

Fiber Basics

- Advantages
- Key Concepts
- Network Components
- Fiber Network Types
- Safety Concerns

© Pearson Technologies 1- Basics 25

25

Network Components

- Fiber
- Cable
- Connectors
- Splices
- Passive devices

© Pearson Technologies 1- Basics 26

26

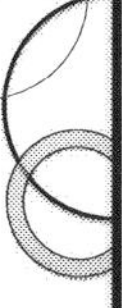

Fiber

- Fiber function is to guide the signal from transmitter to receiver with minimal changes to the signal
- Minimal change means
  - Minimum power loss
  - Minimal signal dispersion
    - Dispersion is distortion of optical signal
    - Excessive dispersion results in signal inaccuracy

© Pearson Technologies 1- Basics 27

27

Cable

- Cable function is to protect the fiber from environmental effects and installation forces
- Major environmental effects include
  - Moisture
    - Moisture resistance provided by powder and tapes or water blocking gel and grease
    - Powder and tapes are super absorbent polymers (SAP)
  - Rodents
    - Rodent resistance provided by corrugated, plastic-coated, stainless-steel armor

© Pearson Technologies 1- Basics 28

28

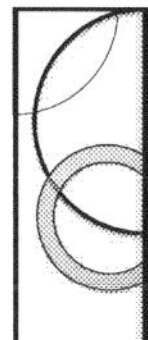

Rodent Resistance

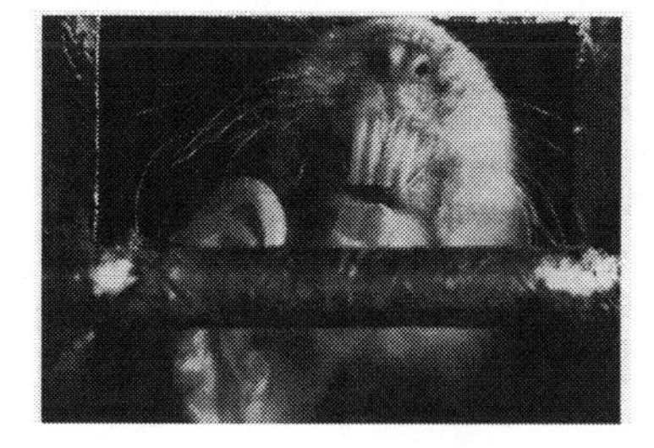

(Courtesy Lucent Technologies)

© Pearson Technologies 1- Basics 29

29

Connectors I

- The connector is a temporary connection
- First function of connector is alignment
  - Of one fiber to another at patch panels
  - Of fiber to transmitter and receiver
    - Alignment results in low power loss
    - Low loss enables delivery of sufficient power to the receiver because
    - Electronics require a minimum signal strength to function properly

© Pearson Technologies 1- Basics 30

30

Connectors 2

- Low loss maintained by dust cap on connector when connector not in use
  - Dust increases connector loss and may result in damage (scratches) to the fiber end
- Second connector function: allow rerouting of signal at patch panels

© Pearson Technologies 1- Basics 31

31

Splice

- The splice is a permanent connection between two fibers
  - Can be semi permanent, e.g., mechanical splice
- Function is fiber alignment to achieve low power loss
- Almost all long, outside plant (OSP), cables are joined (concatenated) with fusion splices

© Pearson Technologies 1- Basics 32

32

Passive Devices

- Manipulate the optical signal in the optical regime
- Used to reduce network cost in FTTH networks

© Pearson Technologies 1- Basics 33

33

Fiber Basics

- Advantages
- Key Concepts
- Network Components
- Fiber Network Types
- Safety Concerns

© Pearson Technologies 1- Basics 34

34

Fiber Networks Can be

- Indoor or outdoor
- Singlemode or multimode

© Pearson Technologies 1- Basics 35

35

Indoor Cables

- Known as 'distribution' and 'premises' cables
- Must meet code requirements for fire retardance
- Usually 50μ, OM3 multimode fiber
- Cable can contain both singlemode and multimode fiber
  - With both fiber types, cables are 'hybrid'
- Used in process control networks to achieve EM and RF immunity of fiber
- Short indoor transmission distance results in splicing being uncommon and
  - OTDR testing of reduced value (See: OTDR dead zone)

© Pearson Technologies 1- Basics 36

36

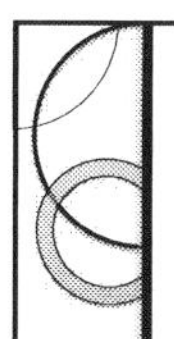

Outdoor Cables

- Fiber is singlemode
  - Multimode use is rare
- Used in telephone backbones, CATV networks, fiber to the home (FTTH) networks
- Are installed in any location
  - Underground conduit
  - Directly buried
  - Aerial suspension
    - Self support (ADSS)
    - Lashed to a messenger wire

 I- Basics 37

37

Fiber Basics

- Advantages
- Key Concepts
- Network Components
- Fiber Network Types
- Safety Concerns

 I- Basics 38

38

Fiber Network Types

- Singlemode
  - Used in almost all outside plant (OSP)
  - Long distance telephone
  - Cell phone tower
  - CATV
  - Premises data
- Multimode
  - Premises data networks (LAN)
  - Process control

 I- Basics 39

39

Fiber Basics

- Advantages
- Key Concepts
- Network Components
- Fiber Network Types
- Safety Concerns

 I- Basics 40

40

Safety Concerns 1

- Eye
  - While creating bare fiber, wear safety glasses
  - Before looking into connector microscope, verify that there is no light in fiber
  - Use a handheld microscope with a built-in IR filter
- Hand
  - As soon as you create a fiber shard, dispose of it

 I- Basics 41

41

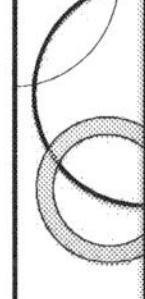

Safety Concerns 2

- Mouth
  - No food, beverage or tobacco in work area
  - In other words, nothing that can find its way to your mouth should be in the fiber work area
- Chemical
  - Before using any chemical, such as fiber cleaner or gel remover, obtain and review its material safety data sheet (MSDS)

 I- Basics 42

42

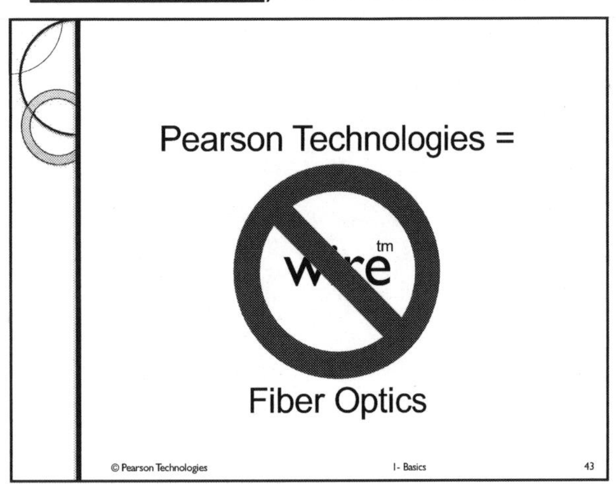
Pearson Technologies =
wire
tm
Fiber Optics
© Pearson Technologies
1- Basics
43

43

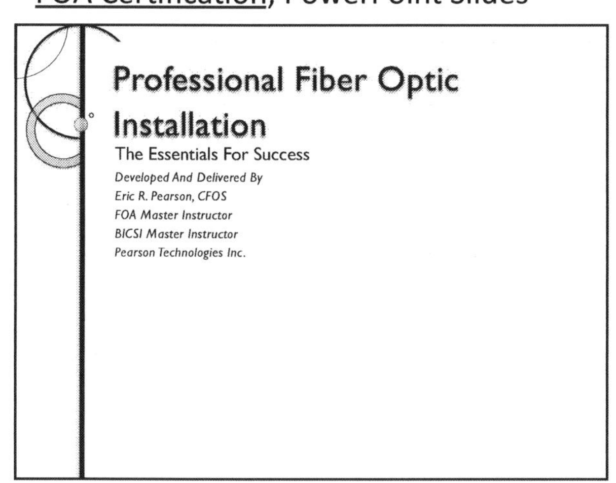

1

Fiber

A Guide to Optical Waveguides

2

Fiber

- Function
- Structure
- Materials
- Glass Fiber Types
- Performance

© Pearson Technologies 4- Fiber 3

3

Function (3-1)

- The fiber is the medium through which light travels
- The fiber's function is to guide the light between the transmitter and receiver with minimum signal distortion
- Minimum signal distortion means
  - Minimum power loss
  - Minimal pulse dispersion

© Pearson Technologies 4- Fiber 4

4

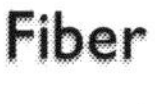

Fiber

- Function
- Structure
- Materials
- Glass Fiber Types
- Performance

© Pearson Technologies 4- Fiber 5

5

Structure-Three Regions

- The fiber provides its function through its structure
- The structure consists of at least two, but usually three, regions
  - Core
  - Cladding
  - Primary coating
- Core and cladding have different
  - Compositions and RIs
    - The difference in compositions confines light to the core

© Pearson Technologies 4- Fiber 6

6

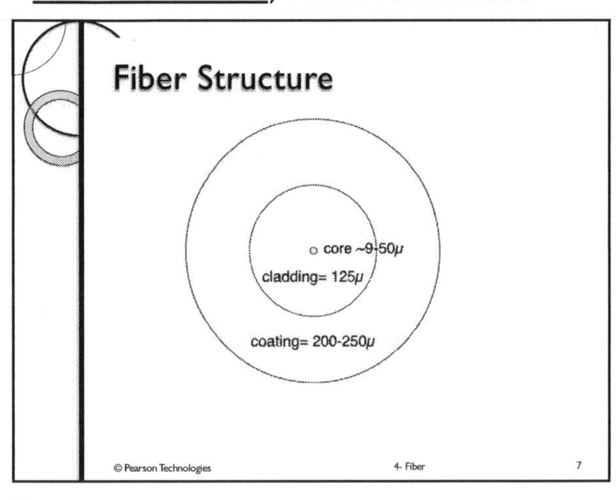

7

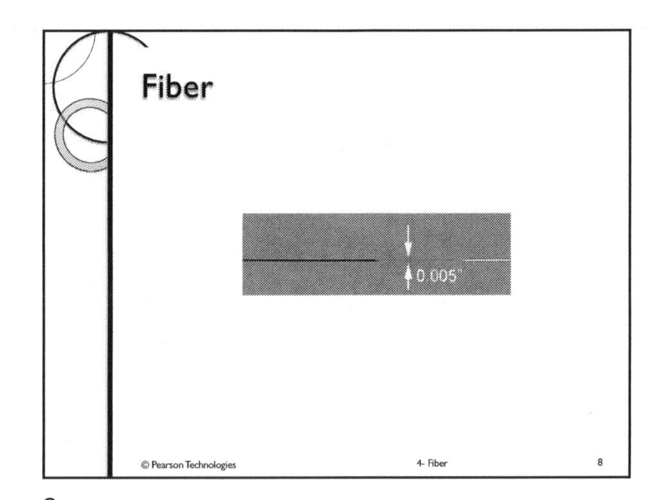

8

Functions of Regions

- The core is the central region of the fiber in which *most* of the light energy travels
- The cladding surrounds the core, confines the light to the core, and increases the fiber size so that it is easily handled
- The primary coating, formerly called the 'buffer coating', protects the cladding from mechanical and chemical attack so that the fiber retains its intrinsic high strength
- A fiber stripper is used to remove the coating for connector installation and splicing

© Pearson Technologies 4- Fiber 9

9

Misconception

- Can cladding be stripped?
  - Never with glass fibers

© Pearson Technologies 4- Fiber 10

10

Diameters (3-1)

- The three regions of the fiber are characterized by their diameters, stated in microns (μ)
- Core diameter
  - The core diameter can range from 8.2μ to 50μ
- Fibers referred to as core diameter / cladding diameter
  - 50/125, 9/125

© Pearson Technologies 4- Fiber 11

11

Core Diameter Terminology

- Fibers have two different central diameter terms
  - Core diameter, for multimode
  - Mode field diameter (MFD), for singlemode
    - MFD ~ core diameter + 1μ

© Pearson Technologies 4- Fiber 12

12

Core Diameter Usage

- Telephone and CATV networks use singlemode fiber with a small mode field diameter (MFD) of ~9µ
- Data networks use fibers with core diameters of 50µ (multimode) or ~9µ (singlemode)

Multimode Status

- Laser optimized (LO), 50µ fiber, OM3, is the current 'de facto' multimode standard
- OM4 is also laser optimized
- OM2 is also 50µ, but not laser optimized

Standards Optimized on 50µ, OM3

- Reason:
  - Increased transmission distance
- 1000BASE-SX: Gigabit Ethernet
- 10GBASE-SX: 10 Gigabit Ethernet
- 40GBASE-SX: 40 Gigabit Ethernet
- 100GBASE-SX: 100 Gigabit Ethernet

Cladding Diameter

- The cladding is always 125µ

Primary Coating

- Primary coating diameter
  - Is usually 245µ or 200µ
    - 245µ is diameter without color coding
    - 250µ is diameter with color coding ink
    - Bend insensitive fibers have 200µ coating
    - Some future fibers will have a 180µ coating
- A fiber stripper is used to remove the coating for connector installation and splicing
- As a practical matter, the coating diameter of little concern to the installer
  - The stripper is sized to the cladding and removes coating of any diameter

Fiber

- Function
- Structure
- Materials
- Glass Fiber Types
- Performance

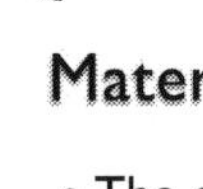

Materials

- The core and cladding can be
  - Glass or plastic
- Fibers can have
  - Glass cores and glass cladding
    - Worldwide, most fibers have glass cores and glass cladding (Our focus)
  - Glass cores and plastic cladding
  - Plastic cores and plastic cladding, aka 'plastic optical fiber' (POF)

© Pearson Technologies 4- Fiber 19

19

POF Characteristics

- Large cores, ~1000µ (1 mm)
- Very high attenuation rates, 100-1000 dB/km
- Step index (SI) profile
- Used in
  - Lighting
  - Short distance, low bit rate, process control links
  - Audio system links

© Pearson Technologies 4- Fiber 20

20

Fiber

- Function
- Structure
- Materials
- Glass Fiber Types
- Performance

© Pearson Technologies 4- Fiber 21

21

3 Types (3-3)

- Type depends on core diameter and structure of core
- Multimode step index (SI)
  - Not used in data communications
- Multimode graded index (GI)
  - Used in data communications
- Singlemode
  - Used in data communications and all outside plant applications

© Pearson Technologies 4- Fiber 22

22

Multimode And Singlemode

- Multimode indicates that rays of light can travel multiple paths through the core
  - These multiple paths are known as multiple 'modes'
- Singlemode indicates that rays of light *behave as though* they are traveling along a single path through the core
  - In other words, light has a single mode in a singlemode fiber

© Pearson Technologies 4- Fiber 23

23

Multimode SI Fibers

- The first type of optical fiber produced was a multimode plastic optical, 'step index' fiber (POF, SI)
- SI fibers have
  - A relatively large core
  - A single chemical composition in core
  - Limited bandwidth
- Their use is not allowed by any of the U.S. data standards

© Pearson Technologies 4- Fiber 24

24

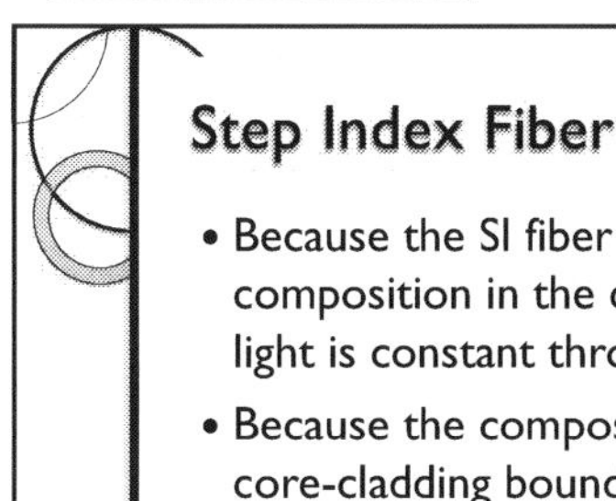

Step Index Fiber

- Because the SI fiber has a single composition in the core, the speed of light is constant throughout the core
- Because the composition changes at the core-cladding boundary, rays of light, known as 'modes' can reflect at that boundary

© Pearson Technologies 4- Fiber 25

25

Step Index Core Profile (3-4)

index of refraction

A B

speed of light

© Pearson Technologies 4- Fiber 26

26

SI Reflection

- Rays of light that enter the end of the fiber at an angle less than or equal to the critical angle will reflect at the cladding boundary
- Rays of light that enter the fiber at an angle greater than the critical angle pass through the boundary and into the cladding

© Pearson Technologies 4- Fiber 27

27

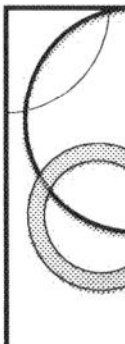

Total Internal Reflection

- Rays of light reflected from one side of the fiber travel to the opposite side
- At the opposite side, the rays reflect again, when the rays remain within the critical angle
- The name of this process is total internal reflection

© Pearson Technologies 4- Fiber 28

28

Total Internal Reflection in SI Fiber

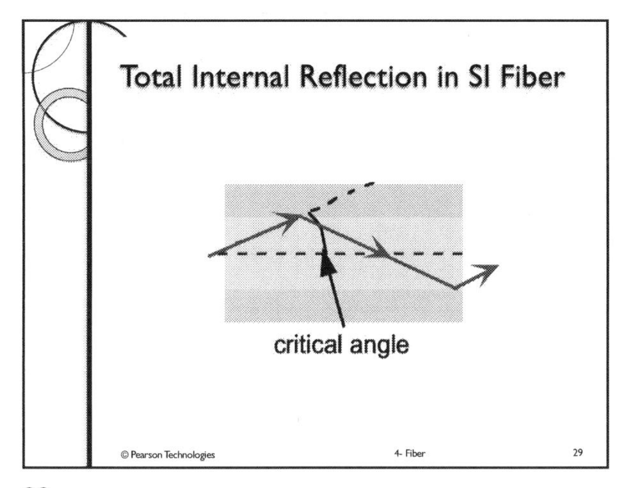

© Pearson Technologies 4- Fiber 29

29

Rays Outside NA Escape

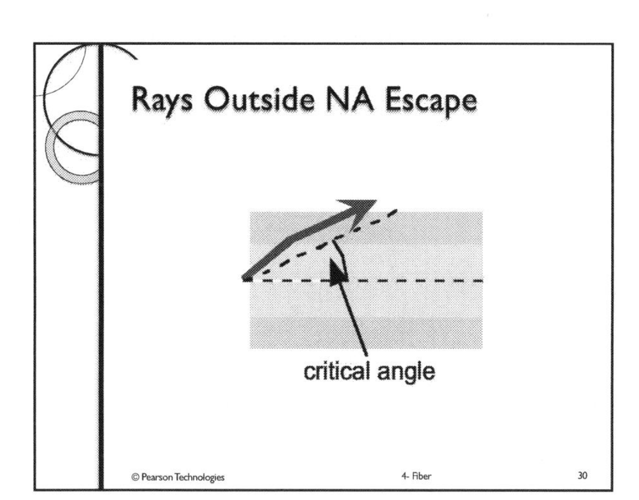

© Pearson Technologies 4- Fiber 30

30

Composition Consequence 1

- Rays of light in the same optical pulse can travel
  - Parallel to the axis of the fiber
  - At any angle to the axis up to a maximum angle defined by the NA of the fiber
- These two rays enter the fiber at the same time, but they travel
  - Different paths and different path lengths

© Pearson Technologies 4- Fiber 31

31

Differing Path Lengths

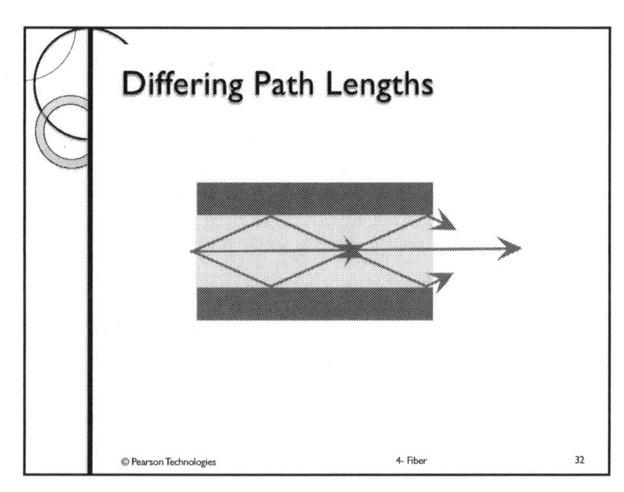

© Pearson Technologies 4- Fiber 32

32

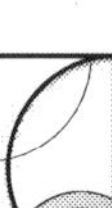

Composition Consequence 2

- Different rays arrive at the end of the fiber at different times
  - Difference in arrival times results in 'dispersion'
  - Dispersion means that each pulse get broader in time as it travels further along a fiber

© Pearson Technologies 4- Fiber 33

33

SI Dispersion Significant

- In SI fibers, dispersion is significant
- Significant dispersion results in a bandwidth that is too low to be of much use
- For this reason, the SI fibers are not used in data networks
- Since SI do not provide the bandwidth required, we needed another fiber type
- That other type is....

© Pearson Technologies 4- Fiber 34

34

Multimode GI Fiber(3-4)

- Graded index (GI) fibers are all glass
- Graded index core has
  - A relatively large core with multiple compositions
  - Multiple compositions result in reduced dispersion
- Core diameter is 50μ
- Graded index (GI) multimode fibers
  - Used at 850nm and 1300nm

© Pearson Technologies 4- Fiber 35

35

GI Core Structure

- The GI fiber is constructed with up to 2500 compositions in the core
- These compositions are chosen so that the speed of light in the center of the core is lowest
  - Or, the RI is highest in the center of the core
- The speed of light in the core increases from the center to the cladding

© Pearson Technologies 4- Fiber 36

36

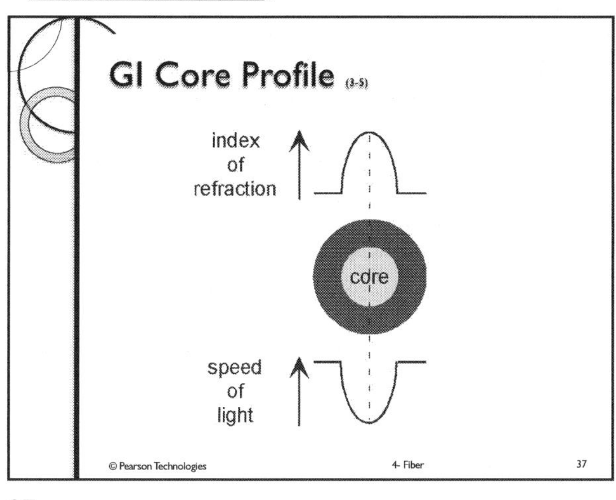

37

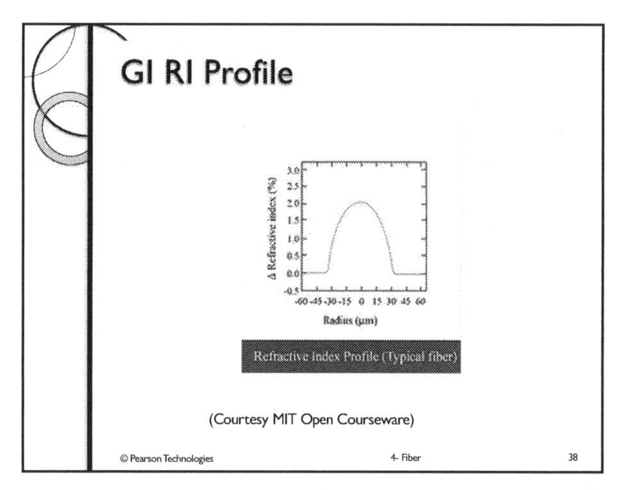

38

GI Compensation

- This core structure compensates for difference in path lengths
- Axial rays at the center travel the shortest path length at the slowest speed
- Critical angle rays travel the longest path length at the highest average speed
- Can you see the compensation?

© Pearson Technologies 4- Fiber 39

39

GI Core Construction

- This compensation allows rays that travel different path lengths to arrive at approximately the same time
- This compensation reduces the amount of dispersion
- For this reason, GI multimode fibers are required for use in data networks

© Pearson Technologies 4- Fiber 40

40

'Curved' GI Paths

- A second consequence of this core structure is a 'bending' (refraction) of the rays
- At each boundary between the layers, the rays bend back towards the axis
- Bending reduces the difference in travel path length of different rays
- In a GI fiber, light travels in paths that appear 'curved'

© Pearson Technologies 4- Fiber 41

41

Ray Paths in GI Fiber

© Pearson Technologies 4- Fiber 42

42

Two 50μ, GI Types

- Standard, as described above
  - Operates at 850nm and 1300nm
- Laser optimized (LO)
  - Laser optimization results in increased transmission distance for ≥ 1 Gbps
  - Core RI profile matched to the characteristics of the light source used at ≥ 1Gbps
  - ≥1 Gbps source is a vertical cavity, surface emitting, laser (VCSEL)
  - VCSELs operate at ~850nm
  - OM3 is the recommended fiber for use with VCSELs

© Pearson Technologies 4- Fiber 43

43

Multimode Dispersion

- Multimode fiber dispersion has two main causes
  - The largest, modal dispersion, results from the different paths taken by different rays (or modes)
  - The second largest, chromatic dispersion, results from the spectral width of the transmitter

© Pearson Technologies 4- Fiber 44

44

Is There An Alternative?

- Two factors, the manufacturing difficulty and the limitation on maximum possible bandwidth, restricted the usefulness of GI fiber and led to a search for another fiber with increased capacity
- That type is….

© Pearson Technologies 4- Fiber 45

45

SINGLEMODE, THE THIRD FIBER TYPE (3-3)

© Pearson Technologies 4- Fiber 46

46

Singlemode (3-5)

- In the 1970's, clever scientists and engineers found such a fiber
- This fiber arose from consideration of the quantum mechanics of light
- This consideration indicated that all rays of light would propagate as though they were traveling parallel to the axis of the fiber.
- Such propagation would result under two conditions: if the core were sufficiently small and the wavelength sufficiently long
- With all rays traveling the same path, multiple path dispersion, modal dispersion, would not occur

© Pearson Technologies 4- Fiber 47

47

'Apparent' Singlemode Paths (3-5)

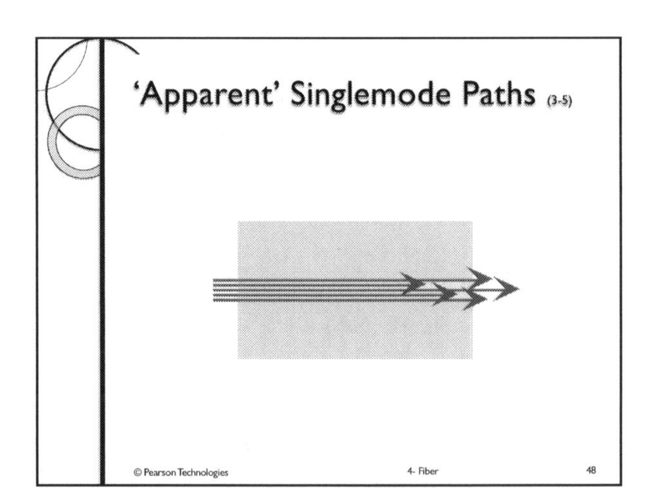

© Pearson Technologies 4- Fiber 48

48

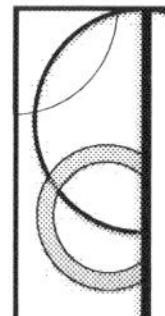

Singlemode Bandwidth

- Since multiple path dispersion is eliminated
  - Bandwidth increases *significantly*
- Since such a fiber would allow a single path, this fiber became known as
  - Singlemode in North America
  - Monomode elsewhere
- Another significant advantage is reduced cost: singlemode fiber cost is much less than that of multimode fiber!

© Pearson Technologies 4- Fiber 49

49

Singlemode Applications

- Singlemode fibers have
  - Essentially unlimited bandwidth
- This high capacity makes this fiber ideal for
  - Long haul, or inter-exchange telephone networks
  - CATV networks
  - FTTH networks

© Pearson Technologies 4- Fiber 50

50

Four Singlemode Characteristics

- Long wavelength operation
- Very small core diameter
- A 'mode field diameter' (MFD)
- Essentially unlimited bandwidth

© Pearson Technologies 4- Fiber 51

51

Wavelength and Core Diameter

- Long wavelength operation
  - 1270-1625 nm
- A 'cut off' wavelength
  - Below 1260 nm wavelength, singlemode fiber behaves as does multimode fiber
- Small core diameter ~9μ

© Pearson Technologies 4- Fiber 52

52

Mode Field Diameter

- Singlemode fibers are characterized by their mode field diameter (MFD)
- The MFD is the diameter within which most of the light energy travels
  - The MFD ~ core diameter + 1μ
  - Example: the Corning Inc. SMF-28 has a core diameter of 8.2μ and a MFD of 9.2μ
  - The generic MFD is 9μ
- Consequence: some optical energy travels in the cladding

© Pearson Technologies 4- Fiber 53

53

Core Diameter and MFD

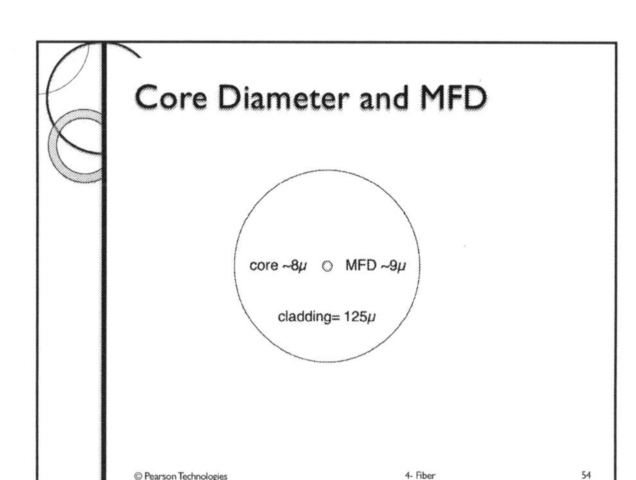

© Pearson Technologies 4- Fiber 54

54

Typical MFDs

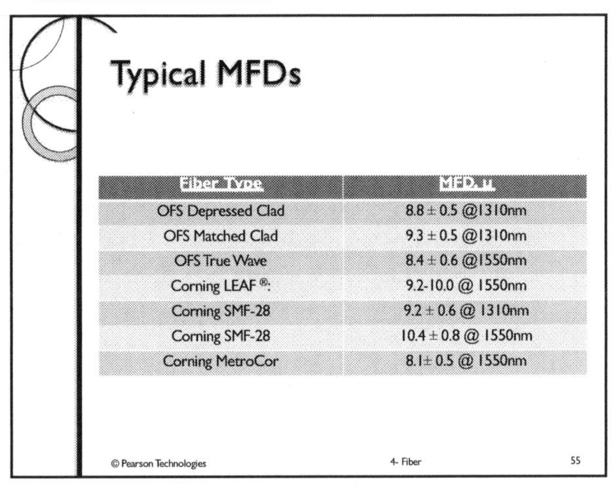

Fiber Type	MFD, μ
OFS Depressed Clad	8.8 ± 0.5 @1310nm
OFS Matched Clad	9.3 ± 0.5 @1310nm
OFS True Wave	8.4 ± 0.6 @1550nm
Corning LEAF ®:	9.2-10.0 @ 1550nm
Corning SMF-28	9.2 ± 0.6 @ 1310nm
Corning SMF-28	10.4 ± 0.8 @ 1550nm
Corning MetroCor	8.1± 0.5 @ 1550nm

© Pearson Technologies 4- Fiber 55

55

Light In Cladding?

- To understand this phenomenon, we step away from fiber optics for a minute.
- Imagine that you are working with a son or daughter on a high school science fair project. The project is to build a model of a single mode fiber.
- You and your son or daughter has decided to use a piece of pipe to simulate the fiber; the center to simulate the core; the wall, the cladding.
- You decide to use a ping-pong ball to model the photon traveling down the core of the fiber. You intend to shoot the ping-pong ball straight down the pipe, simulating singlemode transmission.

© Pearson Technologies 4- Fiber 56

56

Light In Cladding?

- However, light has some of the properties of an energy field. You have not yet modeled the properties of an energy field.
- After further consideration, you decide to rub the ping-pong ball with a piece of fur to create a static charge on the ball. The static charge creates an energy field.
- As the ball travels down the pipe, close to the inside wall, some of the static field travels in the wall.
- Analogously, some of the singlemode optical energy travels in the cladding.

© Pearson Technologies 4- Fiber 57

57

Essentially Unlimited Bandwidth

- Essentially unlimited= 200 Tbps
  - Source: Lucent Technologies, ~2000
- Singlemode dispersion has 4 causes
  - Chromatic dispersion (largest)
  - Waveguide dispersion
  - Material dispersion
  - Polarization dispersion

© Pearson Technologies 4- Fiber 58

58

Fiber

- Function
- Structure
- Materials
- Light in fiber
- Glass Fiber Types
- Performance

© Pearson Technologies 4- Fiber 59

59

Two Fiber Performances

- Light pulses experience only 2 changes as they travel through the fiber
  - Dispersion
    - Data networks do not require dispersion testing because installation error does not increase (degrade) dispersion
    - Dispersion limits multimode distance before power loss
  - Power loss
    - Is the most important testing factor, as installation error can increase power loss
    - Is one of the two important design factors
    - Power loss limits singlemode distance before dispersion

© Pearson Technologies 4- Fiber 60

60

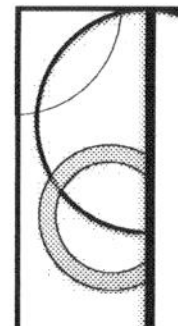

Attenuation Importance (3-11)

- Attenuation, aka power loss, limits transmission distance through reducing power delivered to the receiver
- Excessive attenuation results in insufficient power at the receiver
- Attenuation rate is measured in dB/km
  - Attenuation rate also known as 'attenuation coefficient'
- Attenuation rate is determined by the wavelength and the core size

61

Attenuation Cause

- Rayleigh scattering causes attenuation
  - Scattering is largest source of loss in most links
- Atoms in the core scatter some of the light towards the core boundary
- Some of this scattered light strikes the core cladding boundary at angles greater than the critical angle
- Such light escapes from the core, resulting in attenuation
- Demonstration to follow

62

Rayleigh Scattering

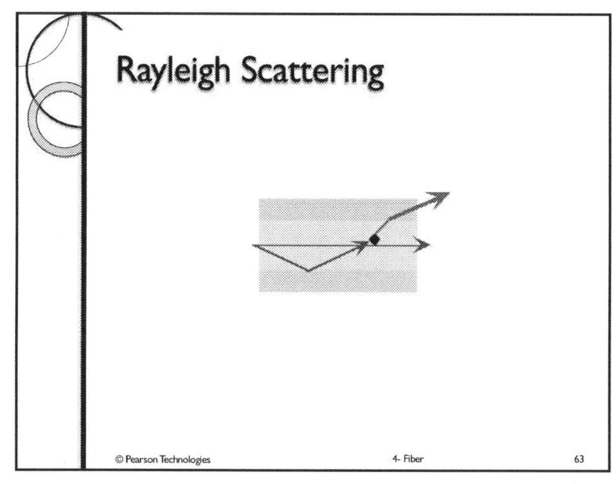

63

Installers Note

- Attenuation is the most important characteristic during installation, since installation mistakes can, and do, result in increased power loss in the cable

64

Attenuation Rates

- The network designer can use the maximum attenuation rate as a cable specification
- The typical attenuation rate is what installers should expect to see on properly designed, manufactured, and installed cables
- The network designer can use both the maximum and typical attenuation rates to calculate the acceptance value, which is the power loss value which the installed cable must not exceed (Ch. 8)

65

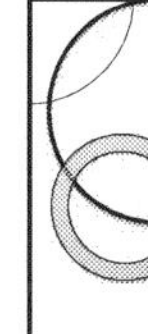

Key Attenuation Fact

- Attenuation rate decreases as wavelength increases (See following examples)

66

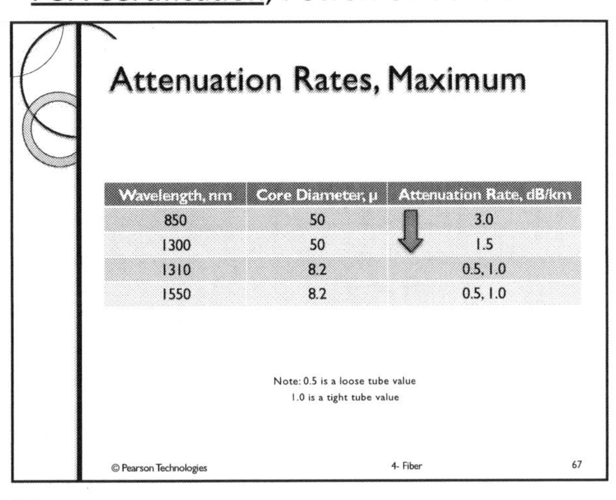

Wavelength, nm	Core Diameter, µ	Attenuation Rate, dB/km
850	50	3.0
1300	50	1.5
1310	8.2	0.5, 1.0
1550	8.2	0.5, 1.0

67

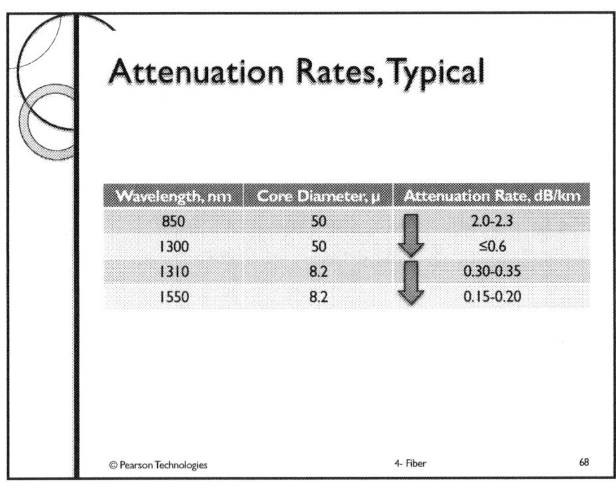

Wavelength, nm	Core Diameter, µ	Attenuation Rate, dB/km
850	50	2.0-2.3
1300	50	≤0.6
1310	8.2	0.30-0.35
1550	8.2	0.15-0.20

68

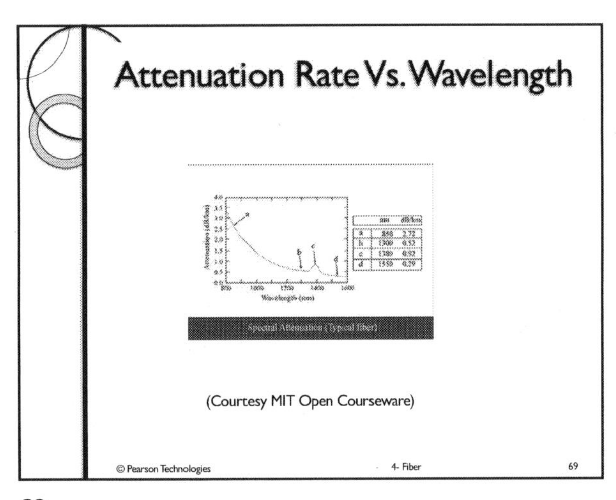

69

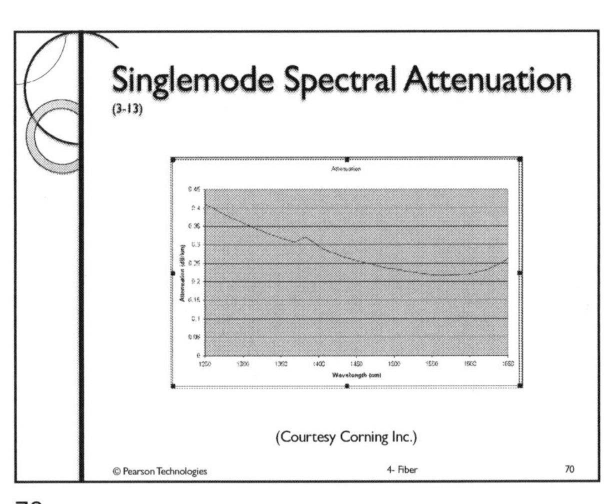

70

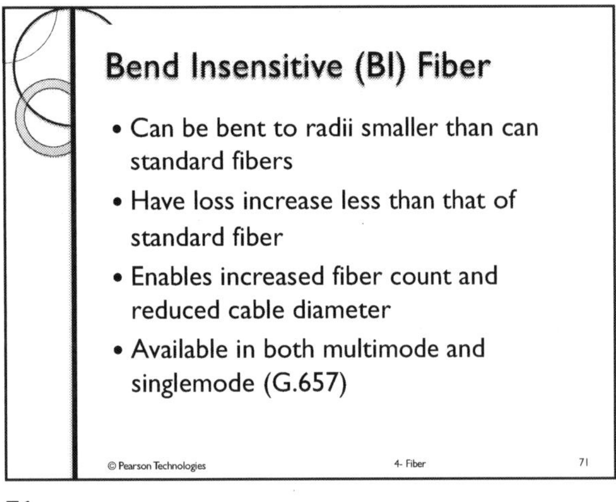

71

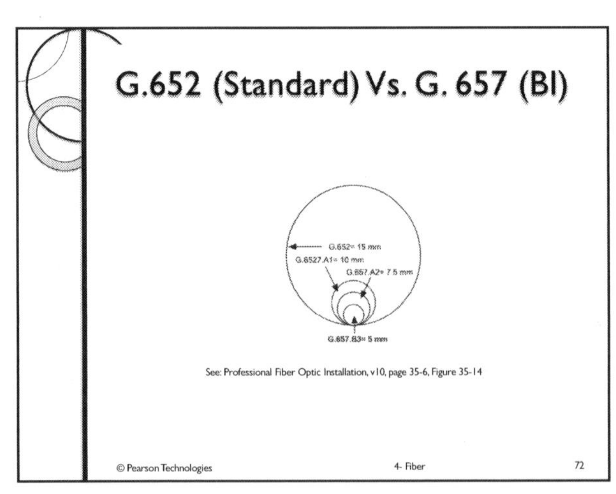

72

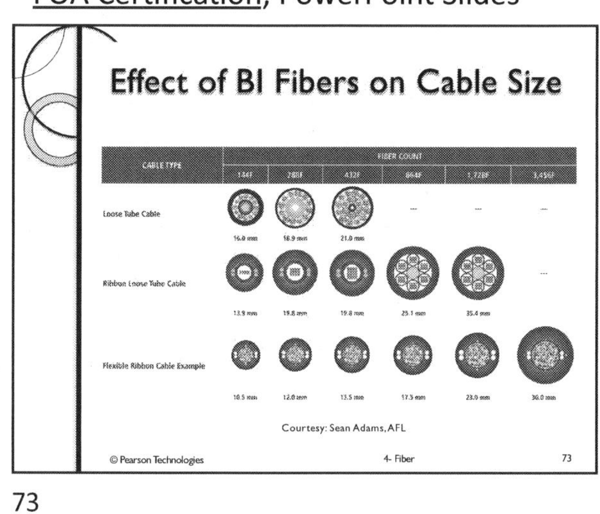
Effect of BI Fibers on Cable Size
CABLE TYPE
FIBER COUNT
144F
288F
432F
864F
1,728F
Loose Tube Cable
16.0 mm
16.9 mm
21.0 mm
Ribbon Loose Tube Cable
13.9 mm
19.8 mm
19.8 mm
25.1 mm
35.4 mm
Flexible Ribbon Cable Example
10.5 mm
12.0 mm
13.5 mm
17.5 mm
23.0 mm
36.0 mm
Courtesy: Sean Adams, AFL
© Pearson Technologies
4- Fiber
73

73

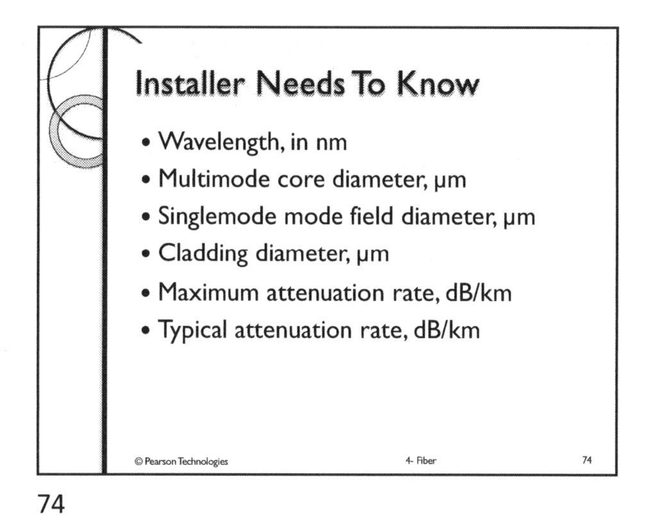
Installer Needs To Know
• Wavelength, in nm
• Multimode core diameter, µm
• Singlemode mode field diameter, µm
• Cladding diameter, µm
• Maximum attenuation rate, dB/km
• Typical attenuation rate, dB/km
© Pearson Technologies
4- Fiber
74

74

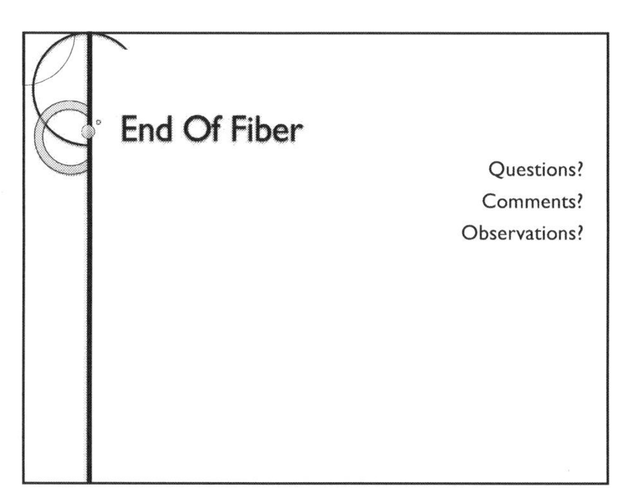
End Of Fiber
Questions?
Comments?
Observations?

75

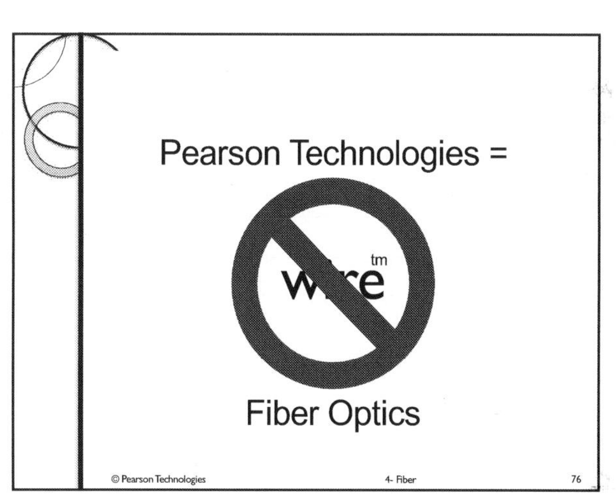
Pearson Technologies =
wire
tm
Fiber Optics
© Pearson Technologies
4- Fiber
76

76

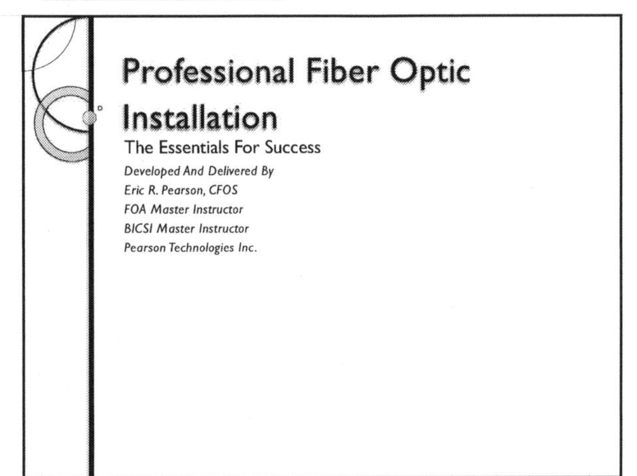

1

Cable

A Package of Protection

2

Cable

- General Information
- Function
- Structure
- Cable Types
- Installation Limitations
- Environmental Requirements

5- Cable 3

3

Terminology

- Common usage
  - A cable containing both multimode and singlemode fibers is a 'hybrid cable
  - A cable containing both fibers and conductors is a 'composite' cable
- Some define these two terms in opposite manner

5- Cable 4

4

Dielectric vs. Conductive Structure

- With the exceptions of armor and steel strength members, all fiber cable materials are non-conductive, i.e., dielectric
- Dielectric designs are preferred
  - The OFN series of cables is an example of indoor dielectric designs
  - Outdoor cables may be, and usually are, dielectric

5- Cable 5

5

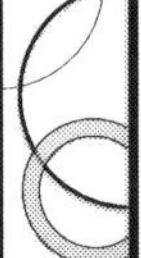

Two Dielectric Advantages

- Improved safety
  - Dielectric designs cannot bring lightning current or ground potential rise into a building
- Reduced installation and maintenance costs
  - Cables with conductive structural elements must be grounded and bonded
  - Grounding is required when cables enter buildings and telecom rooms
  - Bonding is required between cables spliced together (concatenated)

5- Cable 6

6

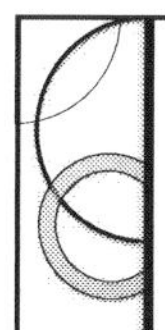

Dielectric Trend

- Safety and cost advantages have created a strong trend towards the use of dielectric, outside plant cables
- Cost advantage has created a strong trend towards the use of dielectric, inside plant cables (OFNR, OFNP)

5- Cable 7

7

Standards 1

- Fiber optic standards are used in designing networks
- The standards define product types and performance requirements
- TIA/EIA-568-E, the Building Wiring Standard defines requirements for both fiber and copper cabling systems
  - TIA/EIA-568-E references
    - ANSI/ICEA S-83-596-2001, Indoor Optical Fiber Cable Standard
    - ANSI/ICEA S-84-640-1999, Outdoor Optical Fiber Cable Standard
    - Color Coding

5- Cable 8

8

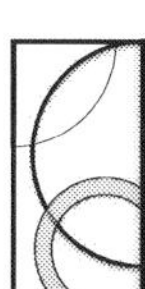

Standards 2

- The -568 logical topology is a hierarchical star

5- Cable 9

9

Cable

- General Information
- Function
- Structure
- Cable Types
- Installation Limitations
- Environmental Requirements

5- Cable 10

10

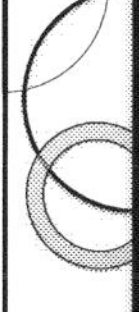

Cable Function 1 (4-1)

- Why cable the fiber?
- Imagine five miles of directly buried glass optical fiber with a diameter of 0.010"!
- While the fiber can easily perform the task of optical communication, it cannot, in and of itself, survive installation and use without additional protection
- The cable is the package that provides protection against fiber damage

5- Cable 11

11

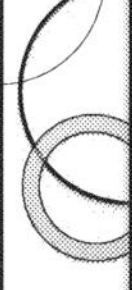

Cable Function 2

- Damage is of at least three forms
  - Breakage
  - Increased loss
  - Reduced reliability
- Damage occurs from
  - Installation forces
  - Environment conditions

5- Cable 12

12

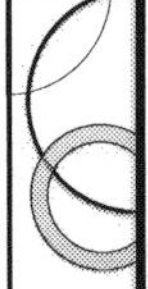

Cable

- General Information
- Function
- Structure
- Cable Types
- Installation Limitations
- Environmental Requirements

5- Cable 13

13

Structure Provides Protection from Damage with

- Buffer tubes
- Water blocking materials
- Strength members
- Binding tapes
- Jacket(s)
- Armor
- Ripcord

5- Cable 14

14

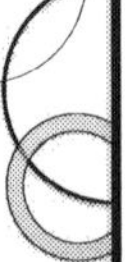

Buffer Tubes

- Definition: the buffer tube is the first layer of plastic placed around a fiber by the cable manufacturer
- Two buffer tube types
  - Loose tube, with a diameter of 2-3 mm or larger
  - Tight tube, with a diameter of 0.9 mm (900µ)

5- Cable 15

15

Loose Buffer Tube

- The inner diameter of the buffer tube is larger than the outer diameter of the fiber
- A loose buffer tube can contain one or more fibers
- 12 is the usual number of fibers
- ≥400 fibers are possible in a single, central, loose tube cable or a ribbon cable with buffer tube diameter of ~ 0.5"

5- Cable 16

16

Loose Buffer Tube Cross Section

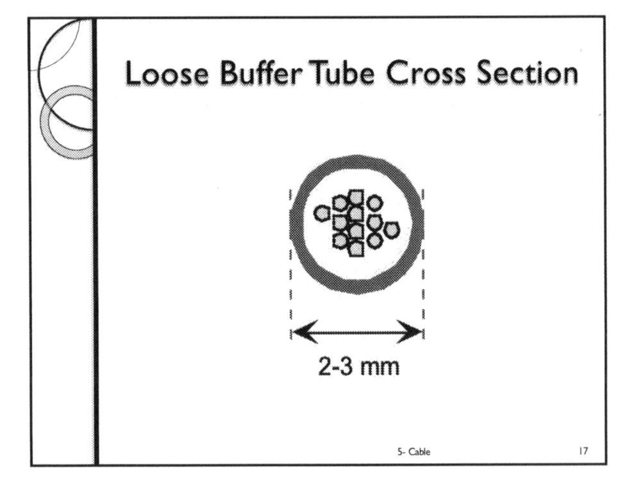

5- Cable 17

17

Loose Buffer Tube Cable

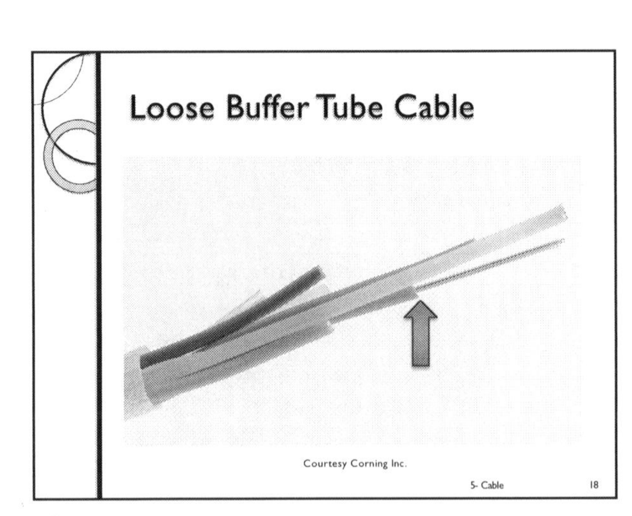

Courtesy Corning Inc.

5- Cable 18

18

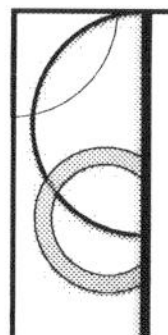

Loose Tube Advantages

- High reliability due to a 'mechanical dead zone'
  - High tension can be required in outside plant installation
  - Dead zone enables very high pulling tension without fiber damage
  - Example
- Relatively low cost
- Ease of achieving moisture resistance
  - Moisture resistance required for cables installed outdoors
- Relatively small size

5- Cable 19

19

Disadvantages

- Reduced flexibility
  - Stiff plastics
- When gel-filled and grease blocked for moisture resistance, increased cost for end preparation
- Increased cost for termination

5- Cable 20

20

Disadvantage 2: Increased Cost

- Results from the the requirement for a fanout tube on each fiber that is to be terminated (connector installed)
- Connectors are not installed directly on 250 μ fiber (insufficient reliability)
- Fanout tube strengthens fiber to withstand repeated handling

5- Cable 21

21

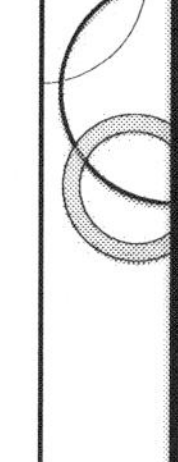

Break Out /Fanout Kit (4-2)

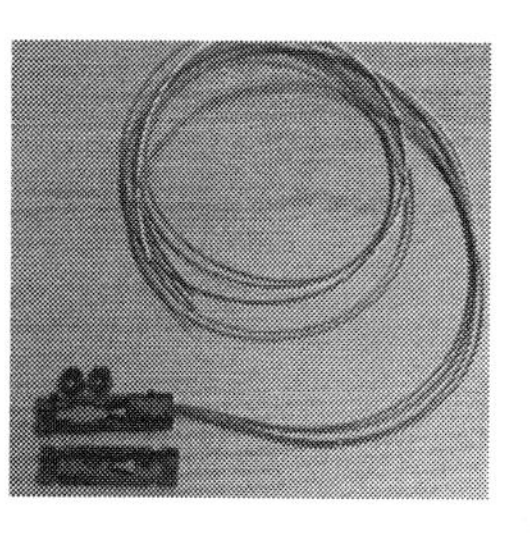

5- Cable 22

22

Fanout Kit

- Also known as
  - Break out kit
  - Furcation kit
  - Splitter kit

5- Cable 23

23

Cost Disadvantage: Increased End Preparation Time

- Example: 0.11 man-hours per fiber per end required to remove gel and grease
- 'Dry water blocked' or superabsorbent polymer (SAP) cables reduce end preparation time by approximately 50%

5- Cable 24

24

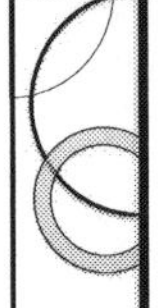

Loose Tube Use Preferred

- Outdoors
- Preference due to
  - Reduced cost for the high fiber counts typical of outdoor cables
  - Ability to place stiff, tough jacketing materials over the loose buffer tubes
  - Increased reliability from 'dead zone'
- It is difficult to put stiff materials around fibers in tight buffer tubes

5- Cable 25

25

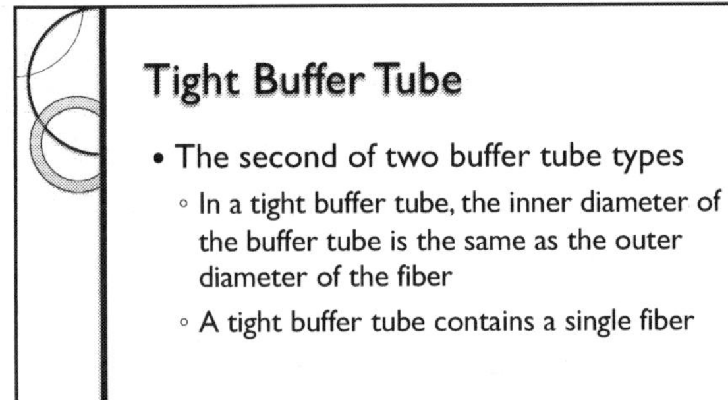

Tight Buffer Tube

- The second of two buffer tube types
  - In a tight buffer tube, the inner diameter of the buffer tube is the same as the outer diameter of the fiber
  - A tight buffer tube contains a single fiber

5- Cable 26

26

Tight Tube

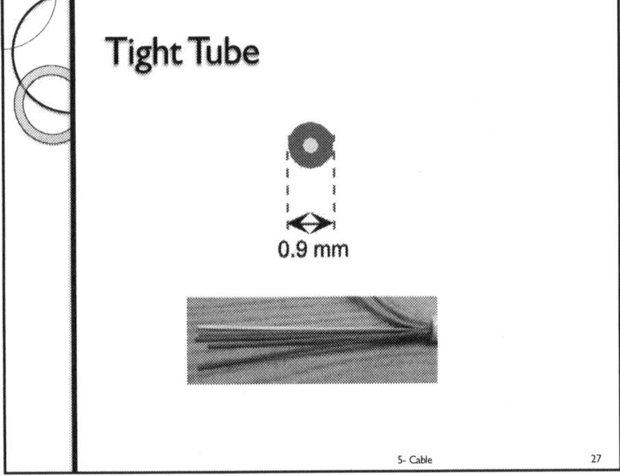

5- Cable 27

27

Tight Tube Advantages

- End alignment in the event of a broken fiber
- Sufficient strengthening of the fiber for termination
- Reduced termination cost: no fanout kit required

5- Cable 28

28

Primary Advantage: End Alignment

- Because there is no space between the outside of the fiber in the inside of the tight buffer tube, broken fiber ends cannot move laterally to create core offset, the major source of power loss in connections
- Thus, a broken fiber in a tight tube cable can provide continuity of signal transmission

5- Cable 29

29

End Alignment

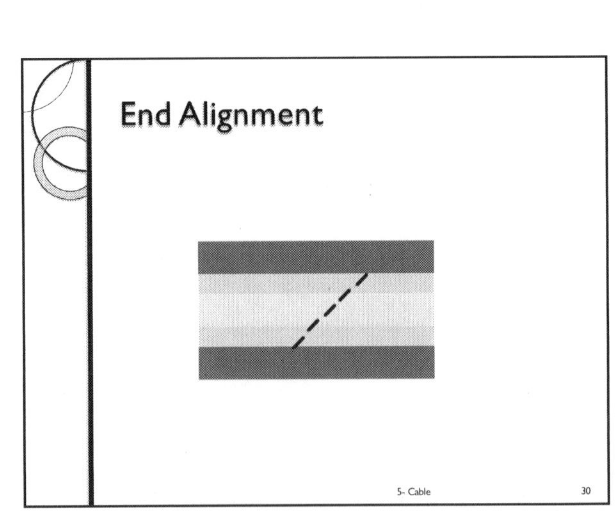

5- Cable 30

30

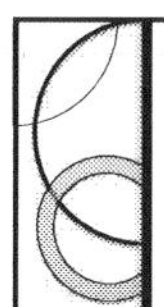

Structure Provides Protection from Damage with

- Buffer tubes
- Water blocking materials
- Strength members
- Binding tapes
- Jacket(s)
- Armor
- Ripcord

5- Cable 31

31

Water Blocking Materials

- All outdoor cables require moisture resistance
- Water blocking materials provide resistance to
  - Cables channeling water into the electronics
  - Fiber breakage from water freezing in cable
  - Degradation of fiber strength from water contacting cladding

5- Cable 32

32

Two Moisture Resistance Systems

- Super-absorbent polymer (SAP) tape and yarn
  - Preferred water blocking system
  - Use reduces labor cost of cable end preparation
- Water blocking gels and greases
  - Rarely used today
  - High labor cost for cable end preparation

5- Cable 33

33

Structure Provides Protection from Damage with

- Buffer tubes
- Water blocking materials
- Strength members
- Binding tapes
- Jacket(s)
- Armor
- Ripcord

5- Cable 34

34

Strength Members Required By

5- Cable 35

35

Strength Members (4-3)

- All cables contain strength members
- These members prevent excessive stretching of fibers during cable installation
- During this installation, the installer attaches the strength members to the pulling rope
- In all-dielectric, self-support cables (ADSS), the strength members allow the cables to withstand high, long-term loads (~2000 lbs-f for 20 years) without fiber breakage

5- Cable 36

36

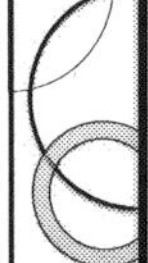

Strength Member Materials

- Aramid yarn, i.e., Kevlar®
- Steel
- Rigid fiberglass epoxy rod
  - Limits bend radius and prevents kinking
  - Provides stable attenuation rate over wide temperature range
- Flexible fiberglass roving

5- Cable 37

37

Strength Member Locations

- In the center of the cable
  - Aka, central strength member
- Outside the buffer tubes and inside the jacket
  - Aka, external strength members
- Between jackets
  - Aka, strengthened jacket
  - A strengthened jacket is used for the all-dielectric, self-support (ADSS) cables used by power utilities between widely-spaced towers

5- Cable 38

38

Central Strength Member

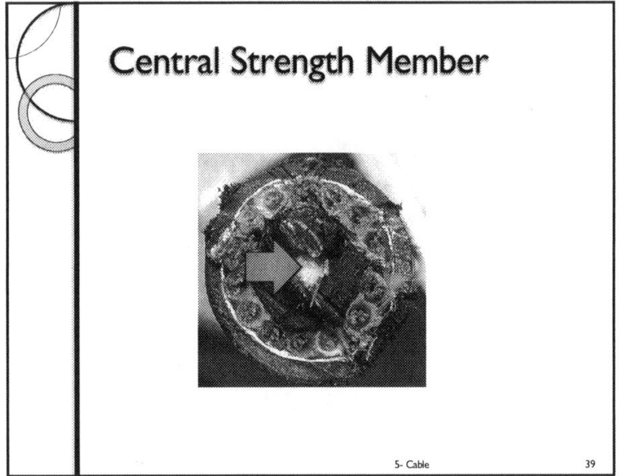

5- Cable 39

39

Central Strength Member

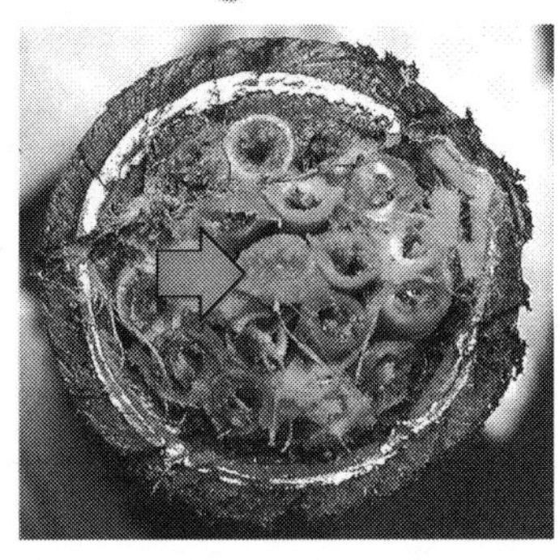

5- Cable 40

40

Structure Provides Protection from Damage with

- Buffer tubes
- Water blocking materials
- Strength members
- Binding tapes
- Jacket(s)
- Armor
- Ripcord

5- Cable 41

41

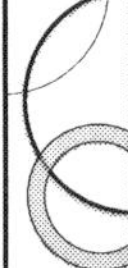

Binding Tapes

- Binding tapes, often Mylar, are placed outside loose buffer tubes to hold the buffer tubes together after the stranding operation and during the jacketing operation
- In addition, such tapes provide a heat barrier so that the heat from the jacket extrusion does not damage the loose buffer tubes
- SAP tape provides additional function of moisture resistance

5- Cable 42

42

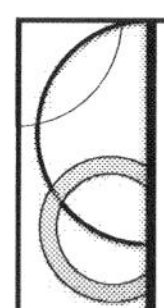

Structure Provides Protection from Damage with

- Buffer tubes
- Water blocking materials
- Strength members
- Binding tapes
- Jacket(s)
- Armor
- Ripcord

5- Cable 43

43

Jackets

- All cables have one or more jackets
- Jackets protect the core of the cable
- Jackets provide protection through resistance to conditions in environment
- Indoor jacket provides compliance with National Electrical Code (NEC)
  - Requires listing by UL
- Common outdoor cable jacket is HDPE, high density polyethylene, which provides moisture and UV resistance

5- Cable 44

44

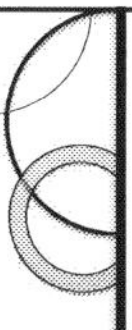

Indoor Jacket Colors

- Yellow: singlemode
- Aqua: LO (50 μ), OM3, OM4
- Lime green: OM5
- Orange: 62.5μ
  - Considered obsolete

5- Cable 45

45

Buffer Tube and Fiber Colors

Blue
Orange
Green
Brown
Slate
White
Red
Black
Yellow
Violet
Rose
Aqua

5- Cable 46

46

Color Code Fibers 1-144

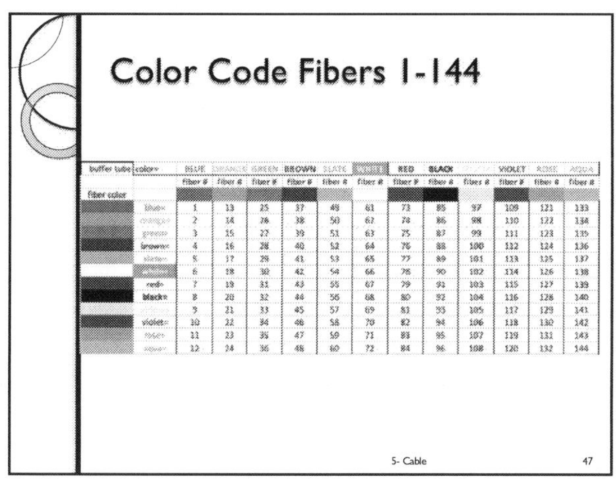

buffer tube	color	BLUE	ORANGE	GREEN	BROWN	SLATE	WHITE	RED	BLACK	YELLOW	VIOLET	ROSE	AQUA
		fiber #	fiber #	fiber #	fiber #	fiber #	fiber #	fiber #	fiber #	fiber #	fiber #	fiber #	fiber #
fiber color													
	blue	1	13	25	37	49	61	73	85	97	109	121	133
	orange	2	14	26	38	50	62	74	86	98	110	122	134
	green	3	15	27	39	51	63	75	87	99	111	123	135
	brown	4	16	28	40	52	64	76	88	100	112	124	136
	slate	5	17	29	41	53	65	77	89	101	113	125	137
	white	6	18	30	42	54	66	78	90	102	114	126	138
	red	7	19	31	43	55	67	79	91	103	115	127	139
	black	8	20	32	44	56	68	80	92	104	116	128	140
	yellow	9	21	33	45	57	69	81	93	105	117	129	141
	violet	10	22	34	46	58	70	82	94	106	118	130	142
	rose	11	23	35	47	59	71	83	95	107	119	131	143
	aqua	12	24	36	48	60	72	84	96	108	120	132	144

5- Cable 47

47

Color Code Fibers 145-288

Add stripe or dash to buffer tube for 145-288

buffer tube	color	BLUE	ORANGE	GREEN	BROWN	SLATE	WHITE	RED	BLACK	YELLOW	VIOLET	ROSE	AQUA
		fiber #	fiber #	fiber #	fiber #	fiber #	fiber #	fiber #	fiber #	fiber #	fiber #	fiber #	fiber #
fiber color													
	blue	145	157	169	181	193	205	217	229	241	253	265	277
	orange	146	158	170	182	194	206	218	230	242	254	266	278
	green	147	159	171	183	195	207	219	231	243	255	267	279
	brown	148	160	172	184	196	208	220	232	244	256	268	280
	slate	149	161	173	185	197	209	221	233	245	257	269	281
	white	150	162	174	186	198	210	222	234	246	258	270	282
	red	151	163	175	187	199	211	223	235	247	259	271	283
	black	152	164	176	188	200	212	224	236	248	260	272	284
	yellow	153	165	177	189	201	213	225	237	249	261	273	285
	violet	154	166	178	190	202	214	226	238	250	262	274	286
	rose	155	167	179	191	203	215	227	239	251	263	275	287
	aqua	156	168	180	192	204	216	228	240	252	264	276	288

5- Cable 48

48

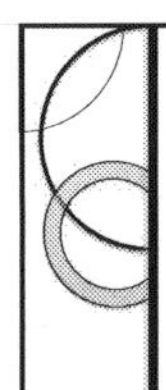

Check Color And Printing: They May Not Agree!

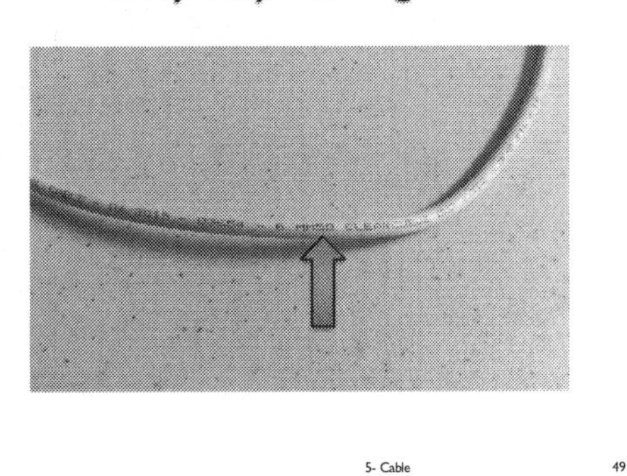

49

Structure Provides Protection from Damage with

- Buffer tubes
- Water blocking materials
- Strength members
- Binding tapes
- Jacket(s)
- Armor
- Ripcord

50

Armor (4-4)

- Corrugated, plastic coated, stainless-steel armor used to prevent
  - Damage from rodent chewing
  - Damage from crush loads
- The corrugation provides some, limited reduction in stiffness
- The plastic coating allows the armor to be heat sealed to itself, providing a measure of moisture resistance
- Stainless-steel does not corrode, so crush and rodent resistances remain constant

51

Rodent Resistance

(Courtesy Lucent Technologies)

52

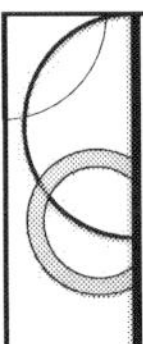

Jacket Over Armor

- Armor always has an external jacket
- Armor may have an internal jacket

53

Armored Cable Without Internal Jacket

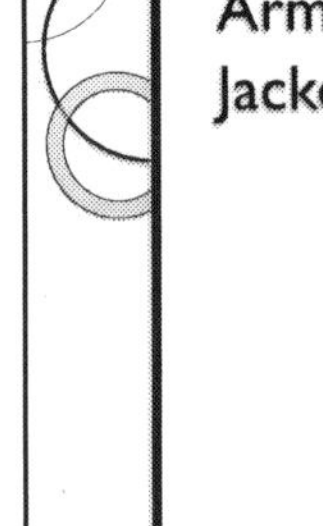

54

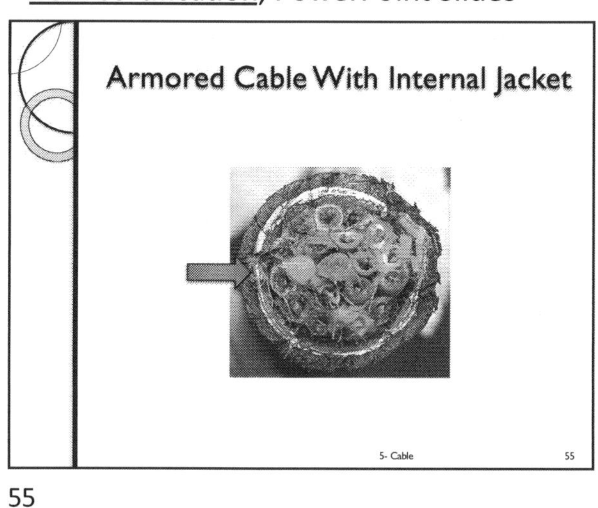

55

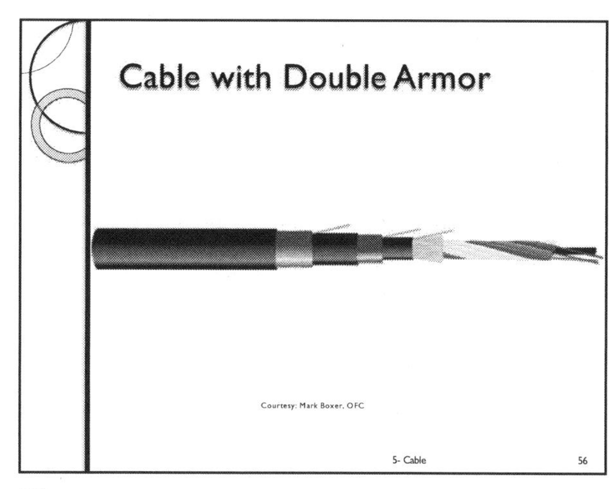

56

Structure Provides Protection from Damage with

- Buffer tubes
- Water blocking materials
- Strength members
- Binding tapes
- Jacket(s)
- Armor
- Ripcord

5- Cable 57

57

Ripcord

- The ripcord protects fibers by eliminating need to use a blade to cut through the jacket
  - Risk: blade will cut fibers
- The ripcord is under jacket or armor
- The ripcord is used to cut through a jacket or armor and eases their removal
- Each jacket and armor requires a ripcord
- Armor can have two ripcords at 180°

5- Cable 58

58

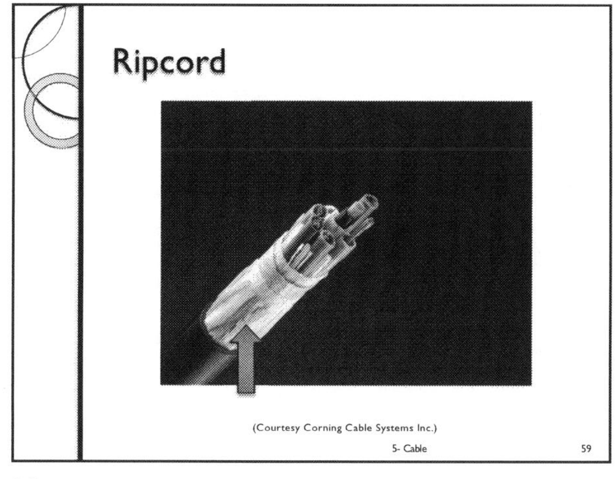

59

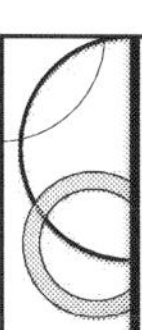

Fillers

- Used as aid to facilitate manufacturing
- Used to create a round cable
  - Round cables are more easily handled than non-round cables

5- Cable 60

60

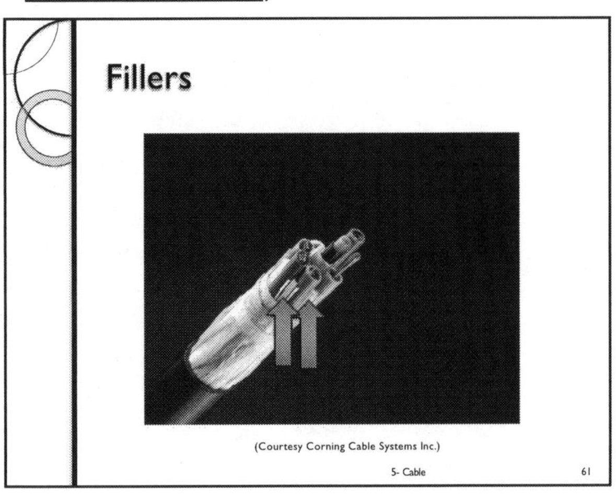

61

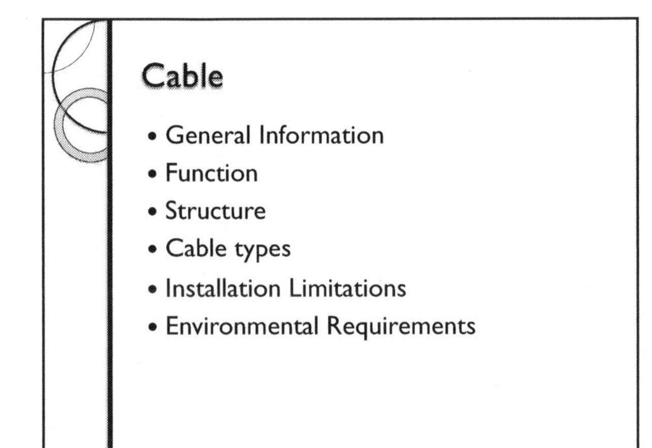

62

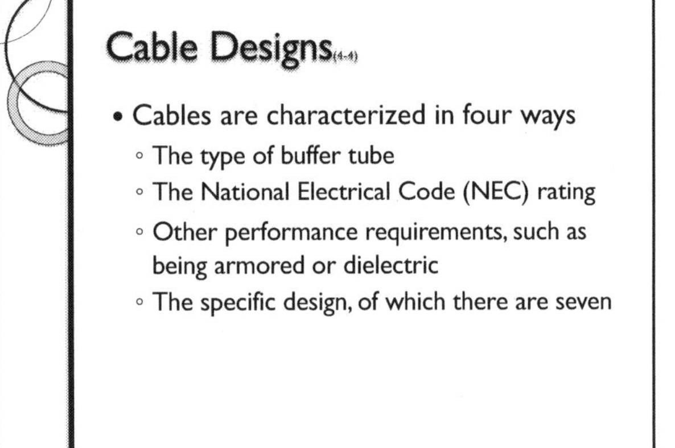

63

Five Common Cable Designs

- Multiple fiber per tube (MFPT)
  - This is author's term, not industry term
- Central buffer tube (CBT)
- Ribbon
- Premises
- Break Out Cable

64

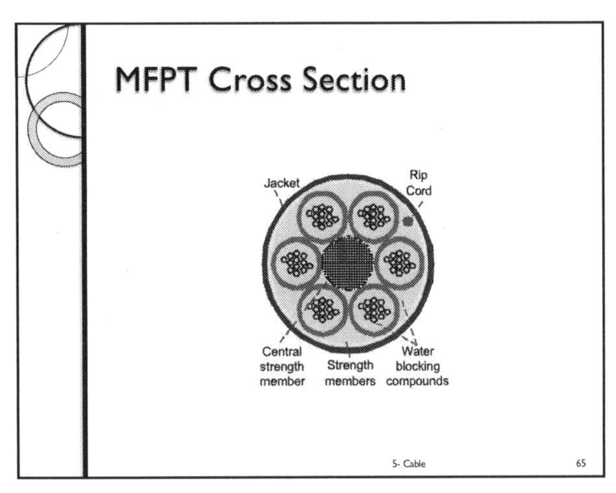

65

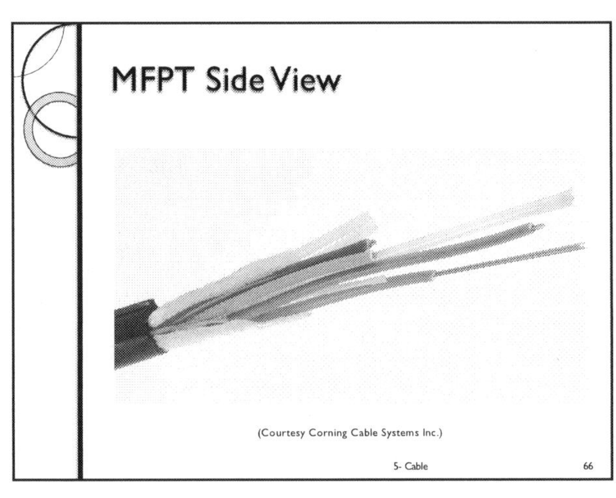

66

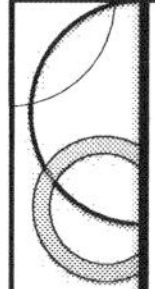

MFPT Description

- The multiple fiber per tube design (MFPT) consists of multiple buffer tubes around a central strength member
- Typical buffer tubes contain 12
- 24 fibers/tube are possible, but not common
- A binding tape surrounds the buffer tubes
- Flexible strength members, either aramid yarn or flexible fiberglass rovings may be over the binding tape and under the jacket
- A ripcord is under the jacket

5- Cable 67

67

Four MFPT Advantages

- Relatively low cost
  - Lower cost/fiber than that of tight tube cable
- Relatively small size
  - Smaller diameter than that of tight tube cable
- Easy mid span access
- Enables high installation load
  - Often required for outdoor installations

5- Cable 68

68

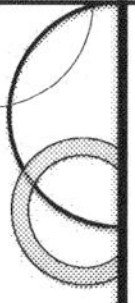

Advantage of Mid Span Access

- Mid span access refers to the situation in which some of the fibers are dropped off between the cable ends
- In this situation, the installer removes the jacket from the cable while still leaving protection, the buffer tubes, on other fibers
- Because of the buffer tubes, it is difficult to damage fibers not terminated at the mid span location

5- Cable 69

69

Advantage of High Load Rating

- In loose tube cables, the fiber is longer than the cable
- When loaded, the cable can stretch without stretching the fiber

5- Cable 70

70

MFPT Disadvantages

- Relatively high labor cost for end preparation
- Enclosures are required on the ends
- Limited flexibility
- Most MFPT cables cannot be used indoors

5- Cable 71

71

Disadvantage: End Preparation Cost

- Must remove water blocking gels and greases
  - Removal of gels and greases is time consuming
  - SAP tape and yarn preferred in place of gels and greases
- Removal of the outer jacket is more difficult than that of tight tube cables

5- Cable 72

72

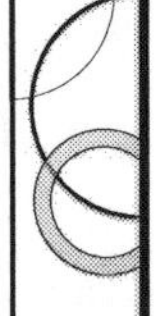

Disadvantage: End Enclosures

- Enclosures increases link costs by a typical value is $3-$6 per fiber per end
- For a 24-fiber cable, the enclosures will add $144-$288 per link

5- Cable 73

73

Disadvantage: Limited Flexibility

- Materials used to meet outside plant environmental requirements make cable relatively inflexible

5- Cable 74

74

Disadvantage: Outdoor Use

- Material used, HDPE, does not meet NEC requirements for indoor cables
  - HDPE is member of paraffin family
- Some MFPT cables can be used indoors, if constructed with NEC-compliant materials

5- Cable 75

75

Common Cable Designs

- Multiple fiber per tube (MFPT)
- Central buffer tube (CBT)
  - This is author's term, not industry term
- Ribbon
- Premises
- Break Out Cable

5- Cable 76

76

Central Loose Buffer Tube Design (CBT)

12 is not the limit!

77

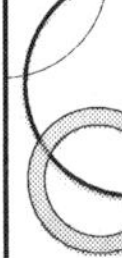

All Fibers In Central Buffer Tube

(4-5)

- There is no reason to limit the number of fibers to 12
- We can place all fibers in a single loose buffer tube
- The buffer tube resides in the center of the cable, resulting in the name 'central loose tube' or 'central buffer tube'
- In this design, fibers are in groups of 12, each group held together by a color-coded thread or yarn
- Up to 216 fibers can reside in the buffer tube
- The buffer tube can include water blocking gel or SAP

5- Cable 78

78

Central Buffer Tube Cross Section

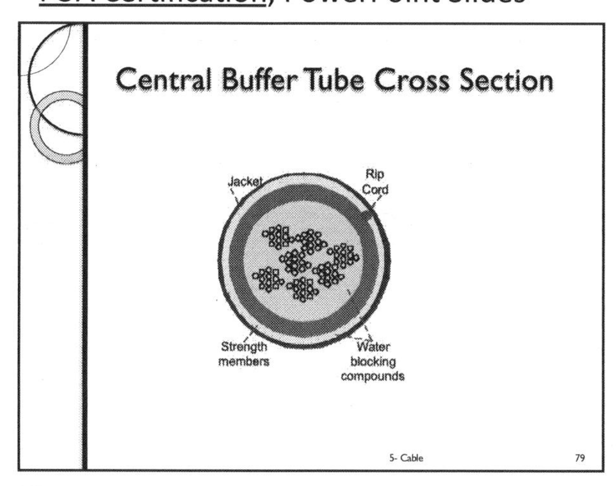

5- Cable 79

79

Central Loose Tube

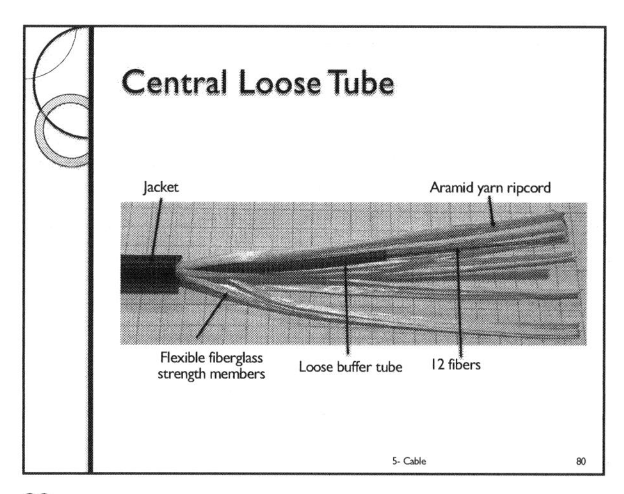

5- Cable 80

80

CBT Structure

- Around the buffer tube are strength members, often flexible aramid yarns or fiberglass rovings
- The empty spaces outside the buffer tube can be filled with water blocking materials
- An outer jacket surrounds the strength members.
- As in the MFPT design, optional additional layers may include additional flexible strength members, armor, or a second jacket

5- Cable 81

81

2 CBT Advantages

- Low-cost
- Reduced cable size

5- Cable 82

82

2 Disadvantages

- Mid span access inconvenient
- Limited to 216 fibers

5- Cable 83

83

Disadvantage: Mid Span Access

- Imagine this design with 216 fibers
- You wish to drop off 12 fibers at a mid span location
- When you remove the jacket and the buffer tube, you expose all 216 fibers
- Murphy's Law states that the fibers you wish to drop at the mid span location are not on the outside of the bundle
- While digging through the bundle to access the desired fibers, you break a fiber
- Of course, Murphy's law states: the fiber you break will not be the one you wanted to terminate at the mid span location
- Mid span access is a major disadvantage

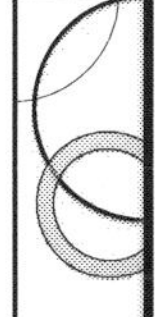

5- Cable 84

84

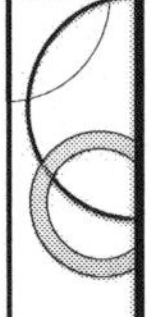

Disadvantage: Fiber Limit

- This design seems to be limited to 216 fibers
- 216 fibers means 18 groups of 12 fibers
- Separation of more than 18 groups is inconvenient during splicing and termination

5- Cable 85

85

Common Cable Designs

- Multiple fiber per tube (MFPT)
- Central buffer tube (CBT)
- Ribbon
- Premises
- Break Out Cable

5- Cable 86

86

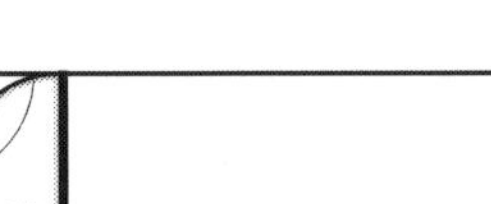

Design 3: Ribbon Cable

Not For Gift Packages!

87

Fiber Ribbon

- A ribbon is a series of 4-24 fibers, precisely aligned and glued to a thin tape substrate
- Ribbon structure enables very high fiber count within a small cable diameter
- Ribbon cables are loose tube designs

5- Cable 88

88

12 Fiber Ribbon, Version 1

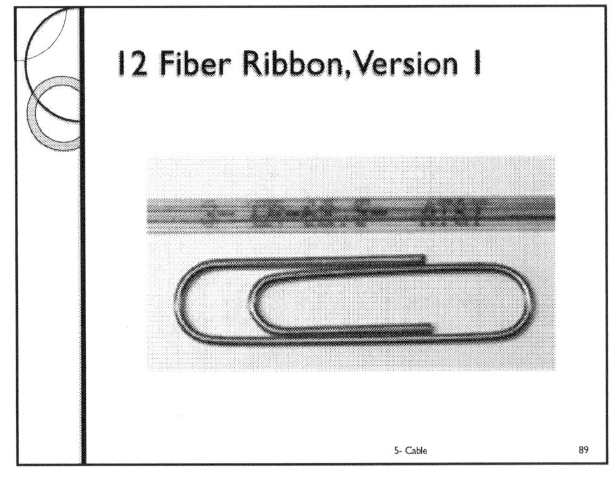

5- Cable 89

89

Ribbon Cable Cross Section

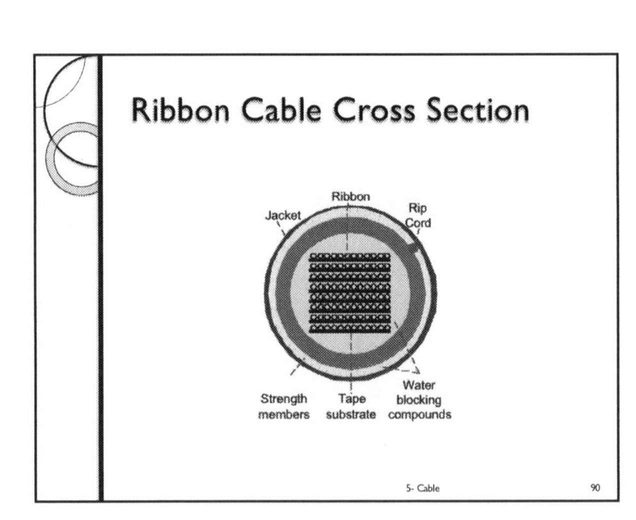

5- Cable 90

90

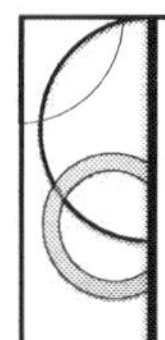

Ribbon Cable, Version 1

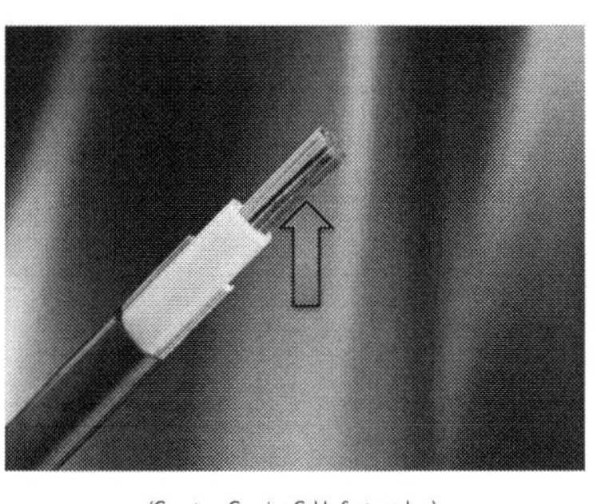

(Courtesy Corning Cable Systems Inc.)

5- Cable 91

91

Ribbon Cable Description

- Ribbons are stacked to create a design with 416 fibers
- The ribbons are enclosed in the central loose buffer tube
- The buffer tube and the rest of the structure are the same as that of the central loose tube design
- An alternative ribbon cable design consists of ribbons stacked in loose buffer tubes, which are stranded around a central strength member, as in the MFPT design

5- Cable 92

92

Ribbon Cable Advantages

- Low cost
  - Ribbon cables have the lowest cost/fiber
- Large fiber count with small diameter
  - Cables with thousands of fibers are always a ribbon design
  - When the fiber count is very high, thousands of fibers, ribbon cables have the smallest diameter and the highest fiber density
  - One champion design: 6912 fibers in 1.46" (37mm) diameter cable
    - (Source: Sumitomo PureAccess™)
- Low splicing cost and time

5- Cable 93

93

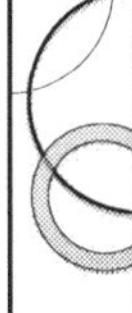

Advantage: Low Splicing Cost

- Pre-alignment of fibers on the tape substrate allows simultaneous fusion splicing of all fibers on the ribbon (aka, mass splicing)
- Mass fusion splicing results in reduced splicing time and cost
  - 12 fiber splicer from multiple sources
  - 24 fiber splicer from Fujikura
- Since all fibers are not individually aligned, ribbon splicing can result in unacceptably high losses on some fibers
- Re-splicing the ribbon is the solution

5- Cable 94

94

Advantage 3: Low Splicing Cost

Why Ribbon Splicing?

Splicing	Ribbon	Single fiber
Fiber count	1728 fibers	1728 fibers
Number of splices	144	1728
Time per splice	~8 minutes	~4 minutes
Cost per splice	~$40	~$25
Total time	~19 hours	~ 115 hours
Total cost	~$6000	~$43,000

Courtesy: Fiber Optic Association
https://www.foa.org/tech/ref/cable/HighFiberCountCables.html

5- Cable 95

95

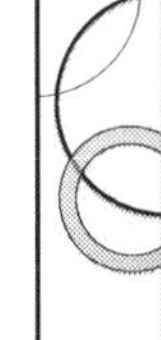

Ribbon Cable, 1728 Fibers

(Courtesy: Ian Gordon Fudge)

5- Cable 96

96

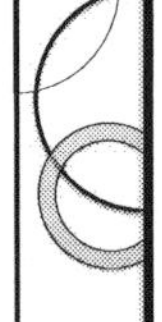

1728 Cable Splicing

- Challenging, to say the least

5- Cable 97

97

High Fiber Count Cable Disadvantage

- Minimum bend radii are larger than the 10x and 20x limits for cables with reduced counts
  - 15x common
  - Manholes and vaults must be larger than those for reduced count cables
- Damaged fibers have significant cost
  - >$50/foot!
  - Hire highly experienced installers only

5- Cable 98

98

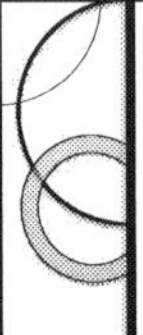

Ribbon Cable, Version 2

- A combination ribbons inside of multiple buffer tubes

5- Cable 99

99

Ribbon Cable, Version 2

(Courtesy Corning Cable Systems Inc.)

5- Cable 100

100

Ribbon Cable: Version 3

- Ribbons formed by fibers with discontinuous attachment
- Produced by AFL Ltd. and Sumitomo Electric Ltd.
- This form enables ribbon to be easily bent in any direction
  - Ribbon Generation 1: does not have this advantage
- The buffer tube and the rest of the structure are the same as that of the central loose tube design
- An alternative ribbon version 3 consists of ribbons stacked in loose buffer tubes, which are stranded around a central strength member, as in the MFPT design

5- Cable 101

101

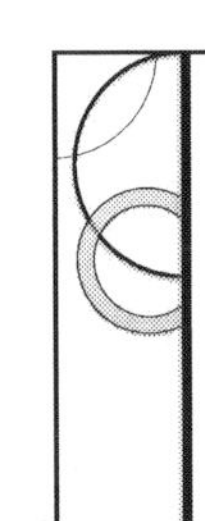

Ribbon for Ribbon Cable, Version 3

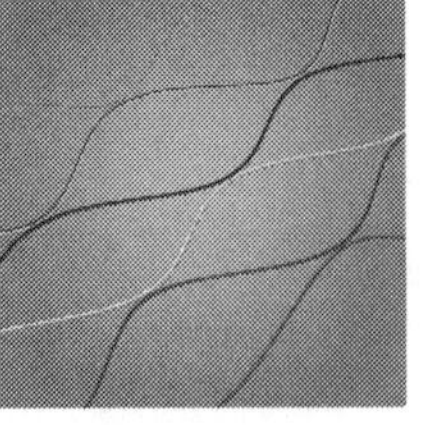

Version 2 Ribbon

Courtesy: AFL Ltd.

5- Cable 102

102

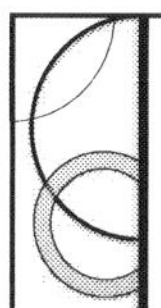

Disadvantages Of Ribbon Cables

- Version 1 ribbons have 250μ fiber spacing
- Version 2 ribbons have 200μ spacing
- Splicing Version 1 to Version 2 somewhat complex
  - Alignment of second-generation ribbons to first generation ribbons require holders designed for different fiber spacings
  - Alignment of second-generation ribbons to first generation ribbons may require 'ribbonizing' 200μ fibers

5- Cable 103

103

Fiber Spacing, 2 Ribbon Generations

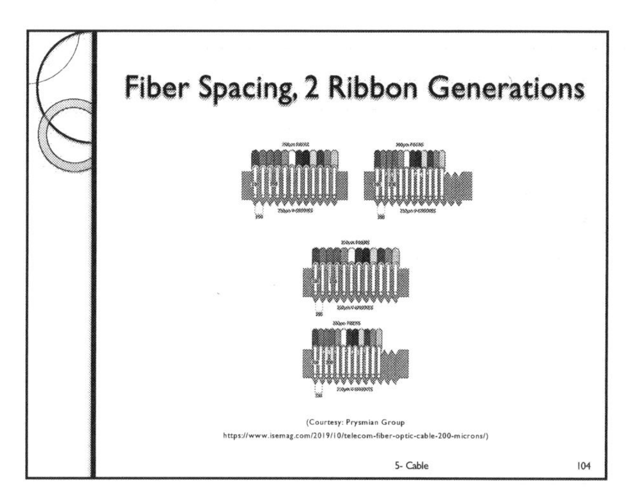

(Courtesy: Prysmian Group https://www.isemag.com/2019/10/telecom-fiber-optic-cable-200-microns/)

5- Cable 104

104

Flexible Ribbon Effect on Size

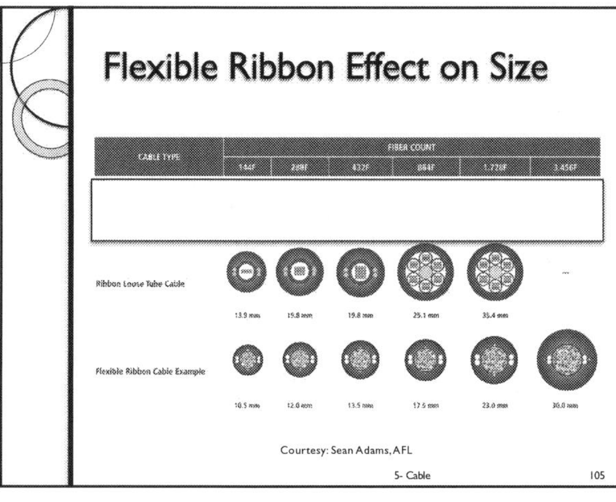

Courtesy: Sean Adams, AFL

5- Cable 105

105

Common Cable Designs

- Multiple fiber per tube (MFPT)
- Central buffer tube (CBT)
- Ribbon
- Premises
- Break Out Cable

5- Cable 106

106

Design 4: Premises Cable (4-7)

- The premises cable, aka 'distribution cable', is the most-commonly used tight buffer tube design
- Commonly used in
  - Indoor networks
  - Field tactical, military applications
- Can be used in outside plant applications

5- Cable 107

107

Premises Cable Structure

- The premises cable structure consists of a centrally located strength member surrounded by a multiple, tightly buffered and stranded fibers
- Around the fibers is a layer of flexible strength members, usually aramid yarns, such as Kevlar®
- A jacket is extruded over the aramid yarns
- For moisture resistance, SAP yarn is placed within the aramid yarns

5- Cable 108

108

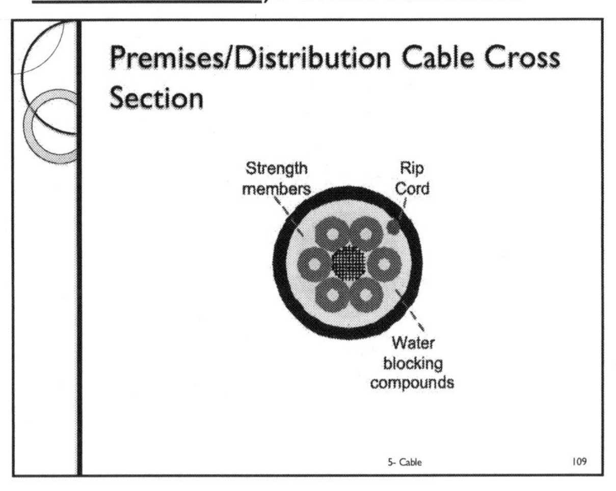

109

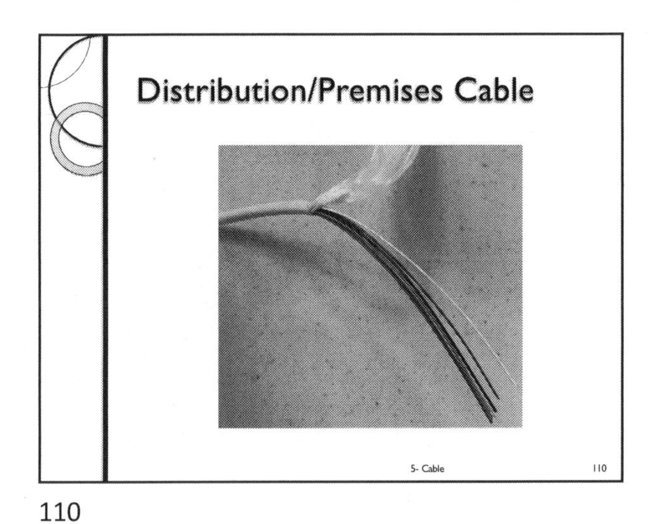

110

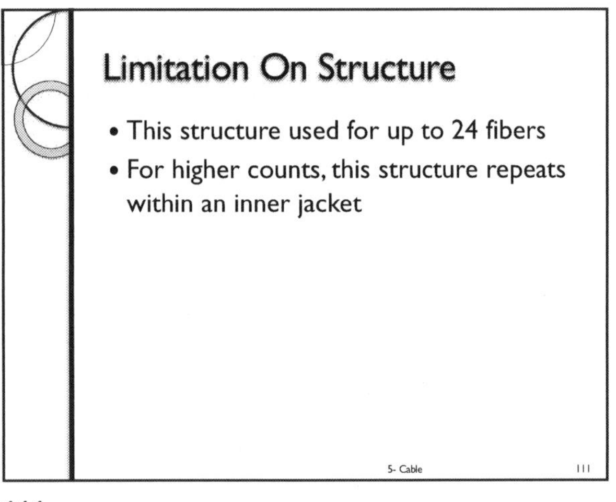

111

High Count Distribution/Premises Cable

5- Cable 112

112

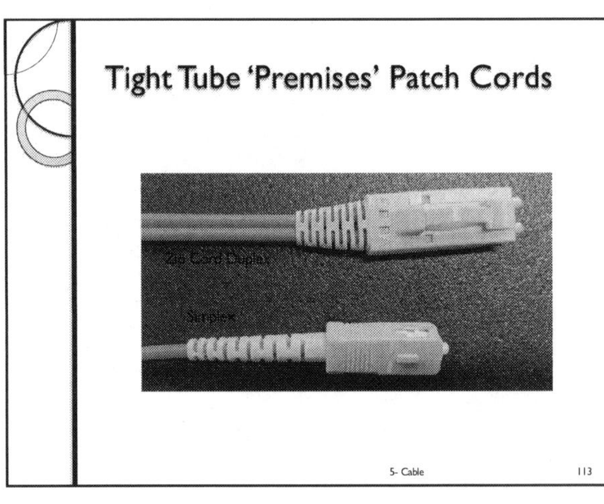

113

Four Premises Cable Advantages

- Low-cost for low to moderate fiber counts
- Low-cost for end preparation
- Low-cost per end for termination
- High flexibility

5- Cable 114

114

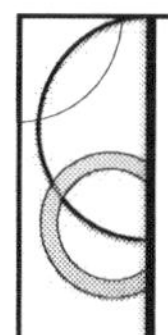

Disadvantage

- Terminations/connectors must be protected by enclosure on end of cable

5- Cable 115

115

Common Cable Designs

- Multiple fiber per tube (MFPT)
- Central buffer tube (CBT)
- Ribbon
- Premises
- Break Out Cable

5- Cable 116

116

Break Out Cable

- Advantage
  - Connectors do not require the protection provided by an enclosure or fiber protection of fanout kit
- Disadvantages
  - Large cable diameter
  - Expensive cable

5- Cable 117

117

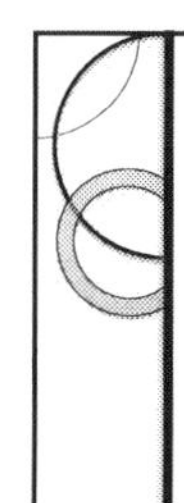

Break Out Cable

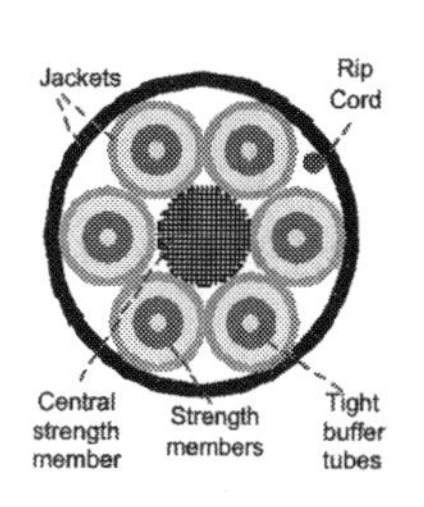

5- Cable 118

118

Break Out Cable

5- Cable 119

119

Cable

- General Information
- Function
- Structure
- Cable types
- Installation Limitations
- Environmental Requirements

5- Cable 120

120

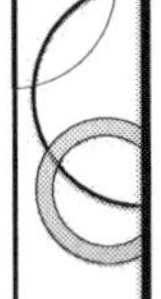

Installation Limitations 1

- Successful installation requires cables to be installed within the limitations of the appropriate cable specifications
- Major installation limitations are
  - Loads
  - Bending

5- Cable 121

121

Installation Limitations 2

- Installation load
- Use load
- Short-term bend radius
- Long-term bend radius
- Storage temperature range
- Installation temperature range

5- Cable 122

122

Excessive Load Causes

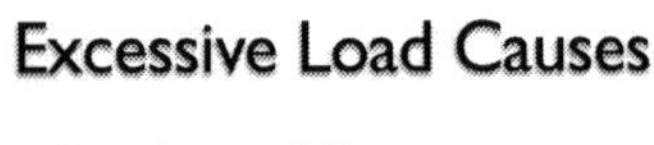

- Breakage of fiber
- Increased attenuation
- Delayed breakage of fiber
  - Delayed breakage is also known as delayed failure and static fatigue
  - Static fatigue is failure of glass materials under relatively low, long-term stress

5- Cable 123

123

Loads

- Short term, or installation, load
  - Short term: maximum load to be applied for short term; i.e., during installation
  - Range: 110-600 lbs.-force (490-2700 N)
  - Important when pulling cable
  - Less important for installation in cable trays
- Long term, or use load
  - Long term: to be left on the cable for entire lifetime
  - Can be, and usually is, zero

5- Cable 124

124

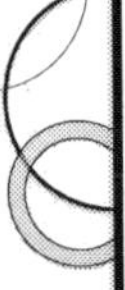

Determine Load Requirement

- From experienced installer
- With software
  - From American Polywater
    - Pull Planner 3000
    - http://www.polywater.com/pullplan.html
  - Greenlee Textron
    - Cable Pull Tension Estimator
    - http://www.greenlee.com/archive/ma5281.pdf

5- Cable 125

125

Use Load

- A use load specification is required if a load is applied to the cable by its environment
- Examples
  - Cables installed between widely spaced poles, buildings, power transmission towers, in risers
- Use loads range widely, depending on the cable type

5- Cable 126

126

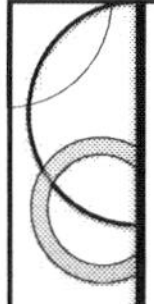

Bending

- Glass does not bend well
- Thus, we have limits on how small a cable can be bent
- Two bend radii
  - Short term, aka installation bend radius
  - Long term, aka use bend radius

5- Cable 127

127

Short Term Bend Radius

- Definition: minimum radius to which cable will be bent while under the maximum recommended installation load
- Aka, installation bend radius
- Always important
- Cable path imposes bend on cable
- Minimum conduit sweep or elbow radius may impose minimum bend radius on cable

5- Cable 128

128

Short-Term Bend Radius

- Rule for minimum short term bend radius
  - ≥ 20 x cable diameter
- Can be critical in manholes
- TIA/EIA-568-E values
  - For inside cable
    - 2-4 fibers: ≥2" (50 mm at 50 lbs.-f/220 N)
    - ≥ 4 fibers: ≥20 x cable OD
  - For inside/outside and outside cable
    - ≥20 x cable OD
  - Use as appropriate

5- Cable 129

129

Long Term Bend Radius

- Definition: minimum radius to which cable will be bent for its lifetime while under no load
- Aka, use bend radius
- Always important
- Cable path imposes bend on cable
- Minimum conduit sweep or elbow radius may impose minimum bend radius on cable
- Size of enclosure used for storage loop imposes minimum bend radius on cable

5- Cable 130

130

Long-Term Bend Radius

- Rule for minimum long term bend radius
  - ≥ 10 x cable diameter
- Important in all locations in which cable is bent
  - Examples: in storage loops, inside pull boxes, manholes, enclosures

5- Cable 131

131

High Count Cables

- Have a long-term minimum bend radius of 15x

5- Cable 132

132

Cable

- General Information
- Function
- Structure
- Cable types
- Installation limitations
- Environmental Requirements

5- Cable 133

133

Environmental Requirements

- Successful installation requires cables to meet the appropriate environmental specifications
- Network designer determines specifications that cables and all other components must meet or exceed
- Major cable environmental specifications are
  - NEC Compliance
  - Moisture resistance
  - Sunlight resistance
  - Operating temperature range

5- Cable 134

134

Environmental Requirements

- Indoor cables
- Outdoor cables

5- Cable 135

135

NEC Compliance

- All cables within buildings need to comply with the requirements of the electrical code
  - The local electrical code, which usually references
    - The National Electrical Code (NEC)
    - NEC requires UL listing

5- Cable 136

136

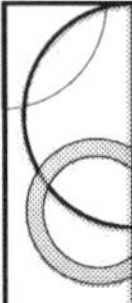

NEC Fiber Cable Classes

- OFNx: a dielectric structure
- OFCx: a conductive structure
  - Such designation applies to fiber cable with conductors for power devices (e.g., cameras, switches)

5- Cable 137

137

NEC Designations

- OFN: for horizontal runs, lowest performance, lowest cost
- OFNR: for vertical runs, increased performance, possibly increased cost
- OFNP: for air handling plenums (i.e., return ducts), highest performance, highest cost

5- Cable 138

138

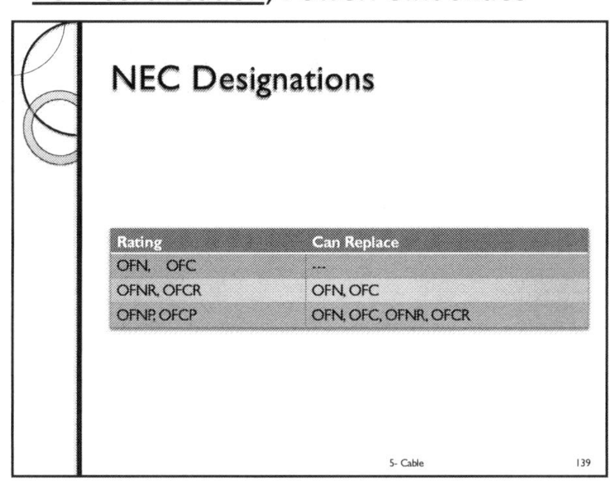

Rating	Can Replace
OFN, OFC	---
OFNR, OFCR	OFN, OFC
OFNP, OFCP	OFN, OFC, OFNR, OFCR

139

Replacements Allowed

- The NEC allows use of an increased rating in an area that requires a reduced rating
- I.e., OFNR for OFN

5- Cable 140

140

Three Trends

- OFN not supplied by many manufacturers
  - They supply OFNR instead
- OFNx supplied
  - Very little use of OFCx (author's opinion)

5- Cable 141

141

Most Common NEC Use

Rating	Can Replace
OFNR	OFN
OFNP	OFN, OFNR

5- Cable 142

142

Environmental Requirements

- Indoor cables
- Outdoor cables

5- Cable 143

143

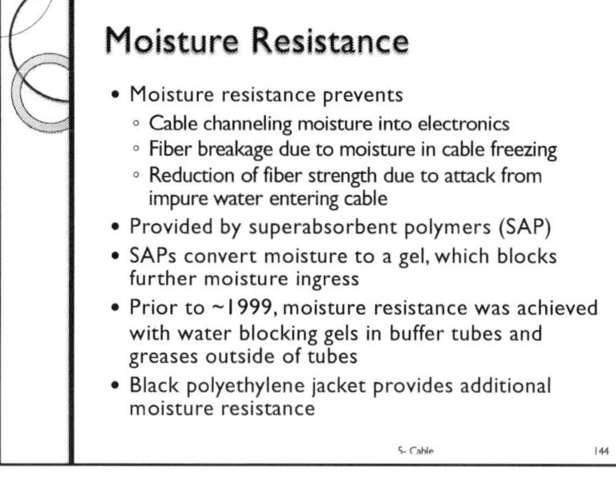

144

Sunlight Resistance

- Provided by black polyethylene
  - Black color from carbon black absorbs UV rays, limiting degradation to thin surface layer of jacket

5- Cable 145

145

Temperature Limits

- The installer must respect two temperature limits
  - Storage temperature range
  - Installation temperature range
- Violation of either of these ranges results in damage to the cable
  - Jackets can crack or be degraded
  - Attenuation rate can increase permanently

5- Cable 146

146

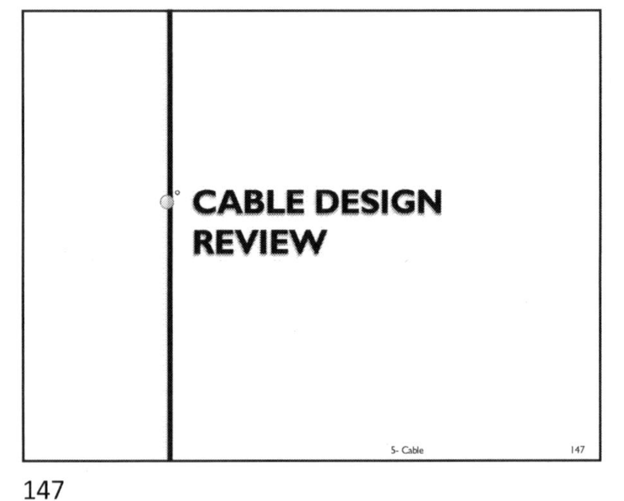

147

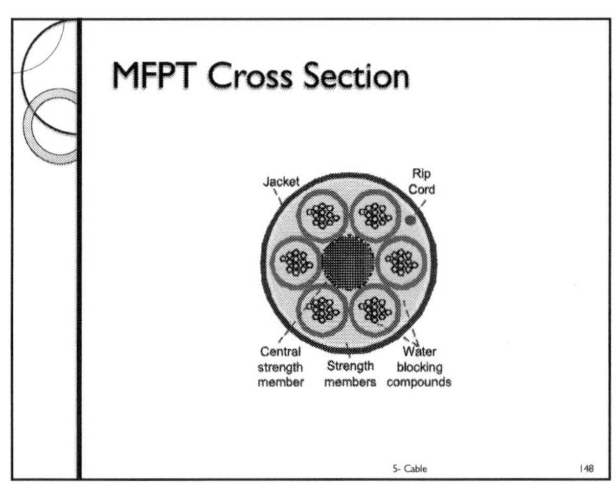

148

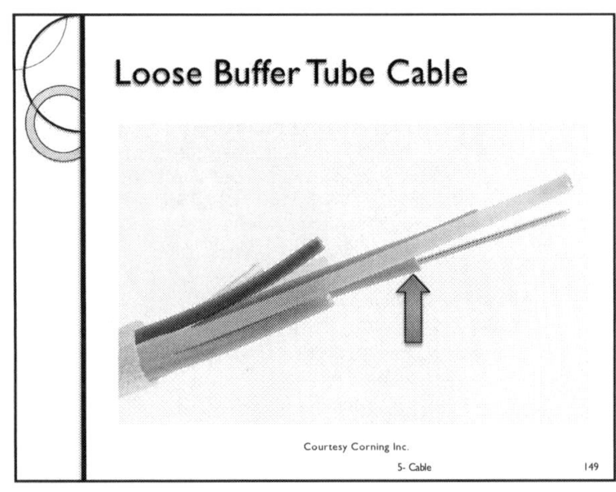

149

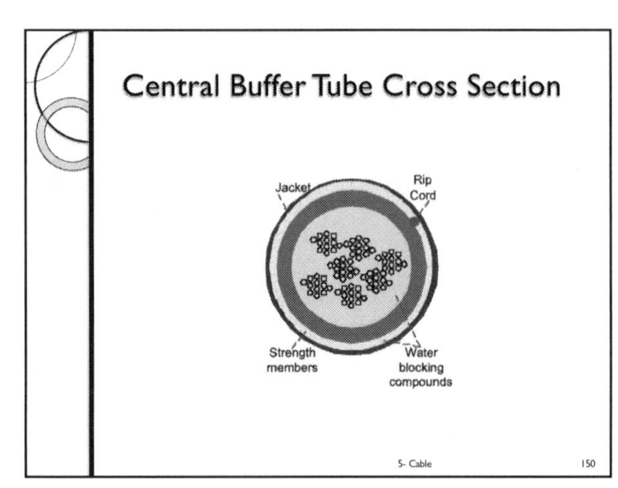

150

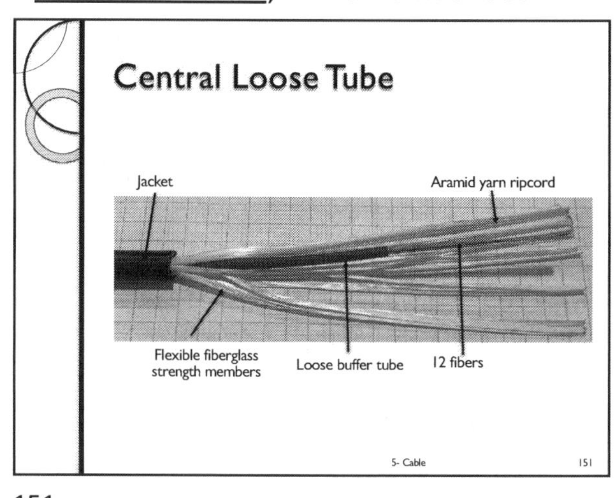

151

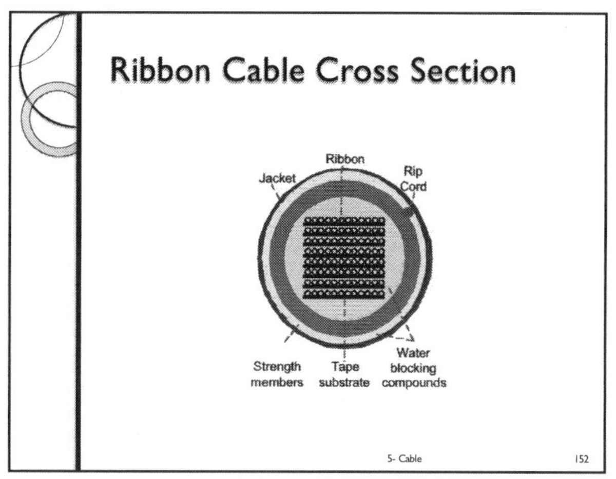

152

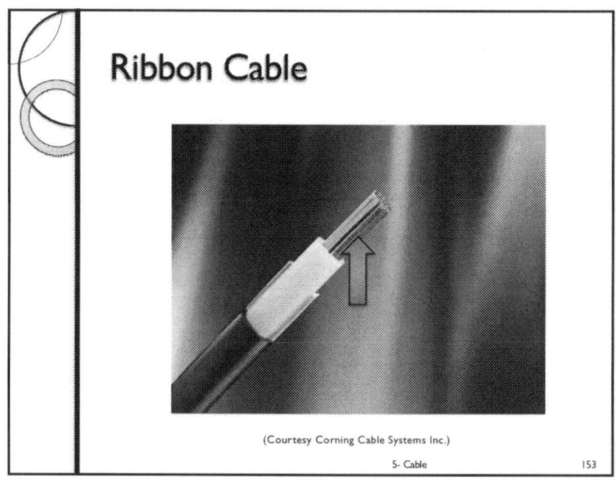

153

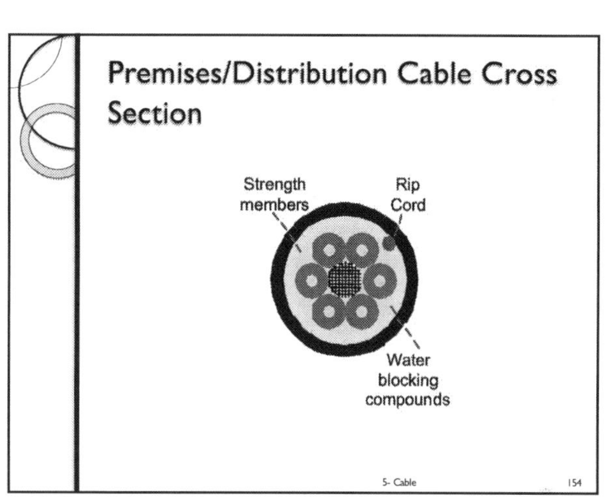

154

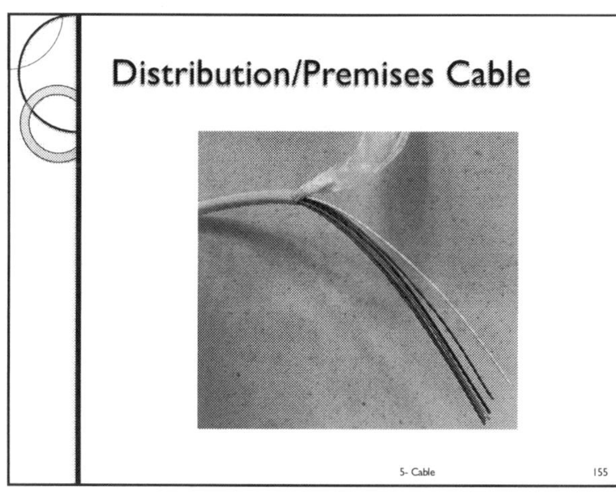

155

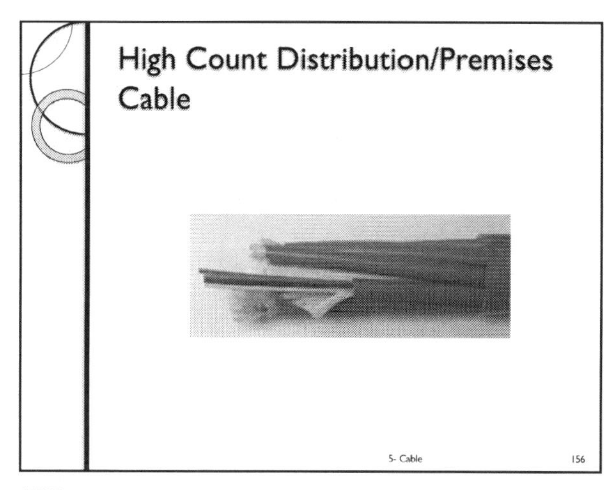

156

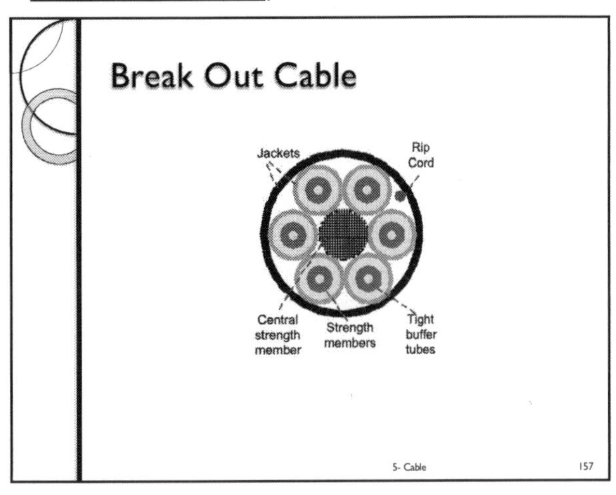

157

Break Out Cable

5- Cable 158

158

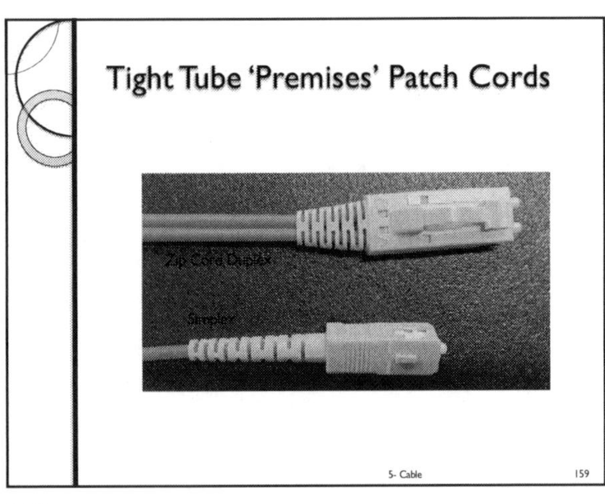

159

Installer Needs to Know

- Short-term, or installation, load
- Long-term, or use, load
- Short term bend radius
- Long term bend radius
- Storage temperature range
- Installation temperature range

5- Cable 160

160

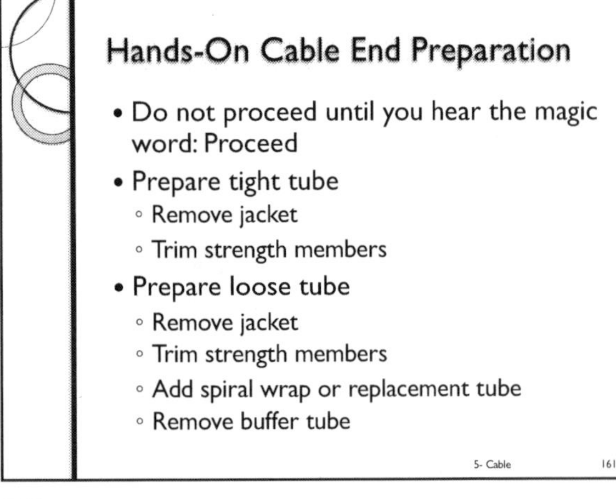

161

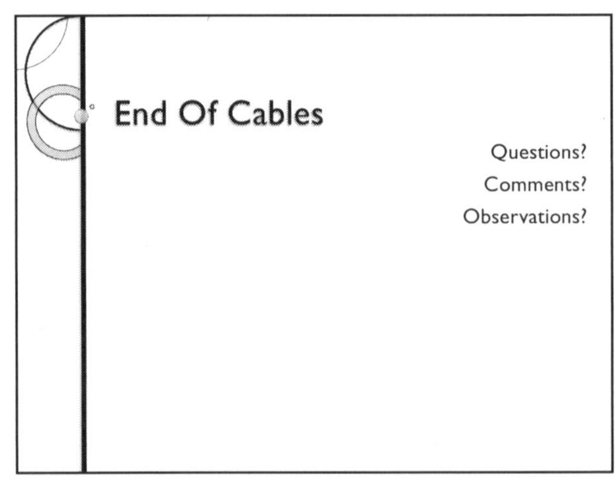

162

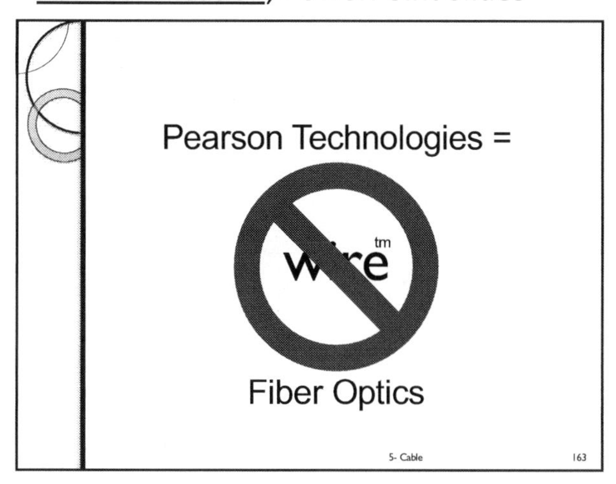

163

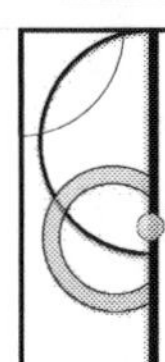

Professional Fiber Optic Installation

The Essentials For Success

Developed And Delivered By
Eric R. Pearson, CFOS
FOA Master Instructor
BICSI Master Instructor
Pearson Technologies Inc.

1

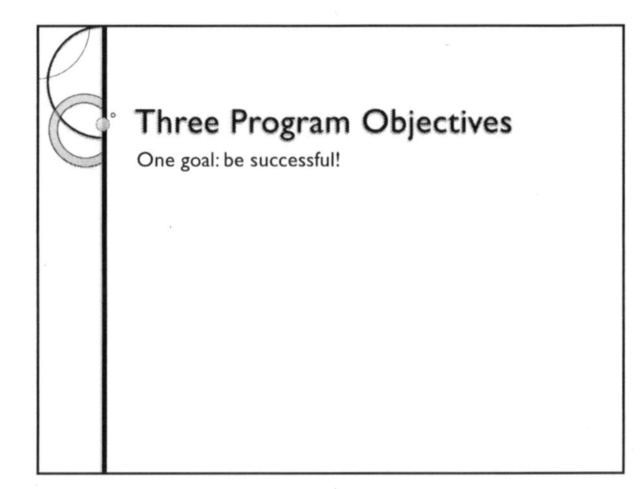

2

Program Objectives

- Success means
  - Low installation cost
  - Low power loss
  - High reliability
- Learn the basics of installation
- Learn the hands-on techniques for success
- Develop basic connector installation, inspection & testing skills
- Pass basic certification examination, Certified Fiber Optic Technician, or CFOT

6- Termination and Splicing 6-3

3

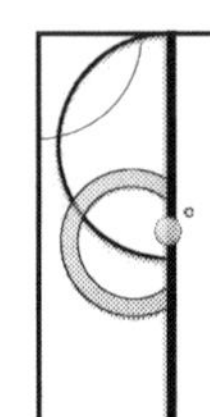

Termination

Beginnings, Concatenations, and Endings

4

Termination

- Means installation of
  - Connectors and
  - Splices
- Connectors create temporary connections
  - They can be disconnected or demounted for signal rerouting
  - As they can do so, durability is a concern
- Splices create permanent connections
  - Splices do not allow rerouting of the signal
  - Durability is irrelevant

6- Termination and Splicing 6-5

5

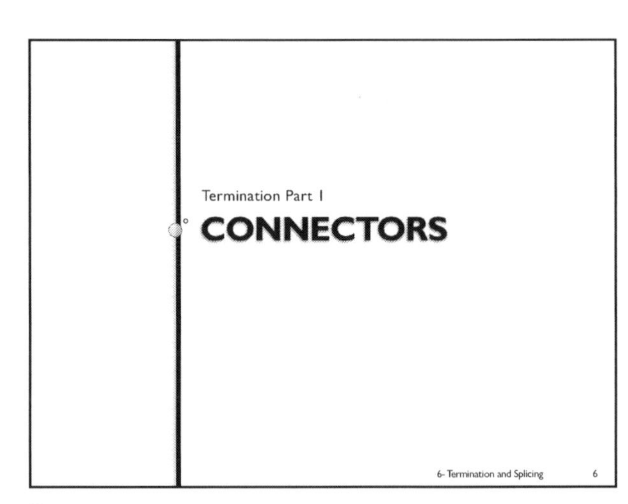

6

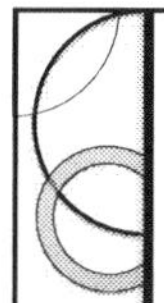

Connectors

- Basic Information
- Key Facts
- Function
- Structure
- Colors
- Performance
- Types
- Installation methods

6- Termination and Splicing 6-7

7

Locations Used

- Anywhere disconnection is necessary
- Anywhere permanent connections are not required
- Anywhere splices are not used

6- Termination and Splicing 6-8

8

Compliance with Standards

- TIA/EIA-568, the Building Wiring Standard, allows use of any connector type with an approved Fiber Optic Connector Intermatability Standard (FOCIS)
- The first version of TIA/EIA-568 required use of the SC
- LC and ST™-compatible have FOCIS documents and comply with the standard

6- Termination and Splicing 6-9

9

More fibers in less space.

A SHORT HISTORY OF CONNECTORS

6- Termination and Splicing 10

10

12 Fibers (2010) Vs. 2 Fibers (1988)

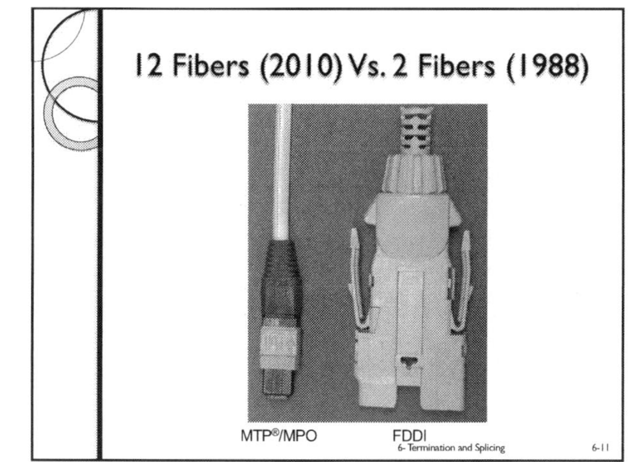

MTP®/MPO FDDI

6- Termination and Splicing 6-11

11

24 Fibers (2011) Vs. 2 Fibers (1988)

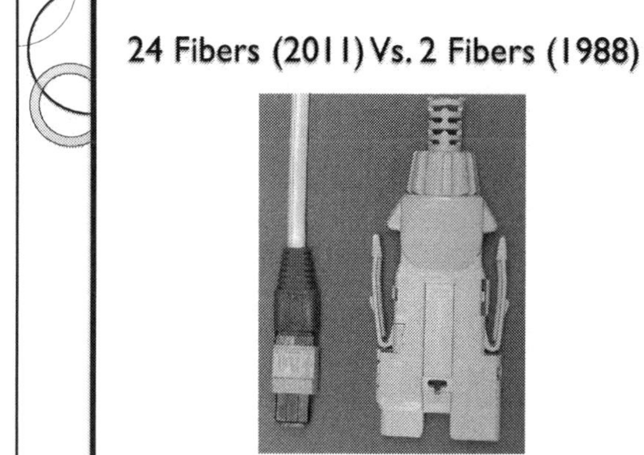

6- Termination and Splicing 6-12

12

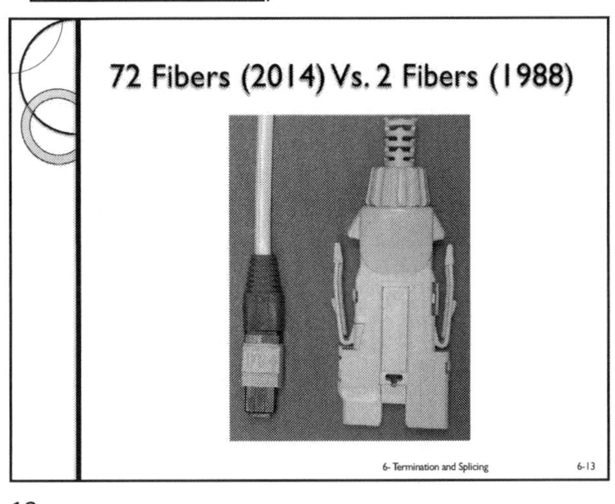

13

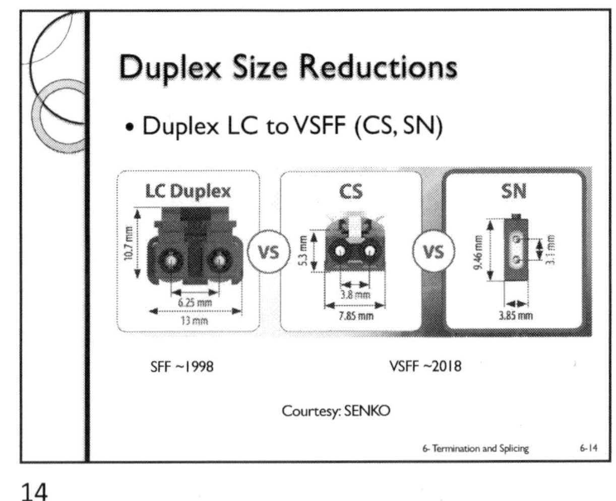

14

Connectors

- Basic Information
- Key Facts
- Function
- Structure
- Colors
- Performance
- Types
- Installation methods

6- Termination and Splicing 6-15

15

Key Facts

- All connectors have keys
  - To provide constant loss
- All current generation connectors are 'push on, pull off'
- All currently used connectors have ferrule diameters of 1.25mm or 2.5mm
- All connectors make physical contact, designated with acronym 'PC'
  - When compared with flat-ended legacy connectors, PC connector ends reduce both loss and reflectance

6- Termination and Splicing 6-16

16

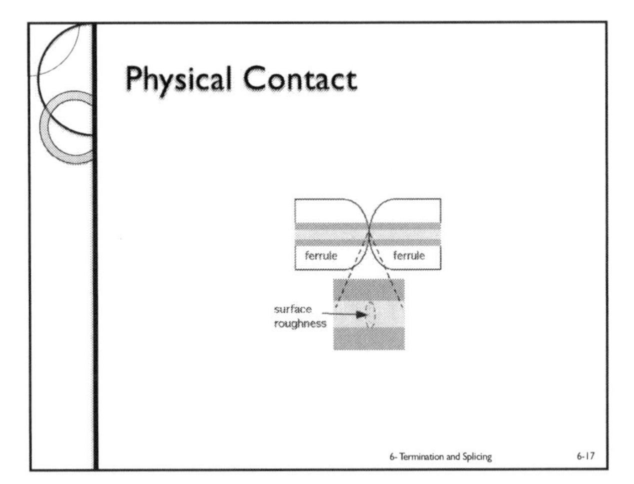

17

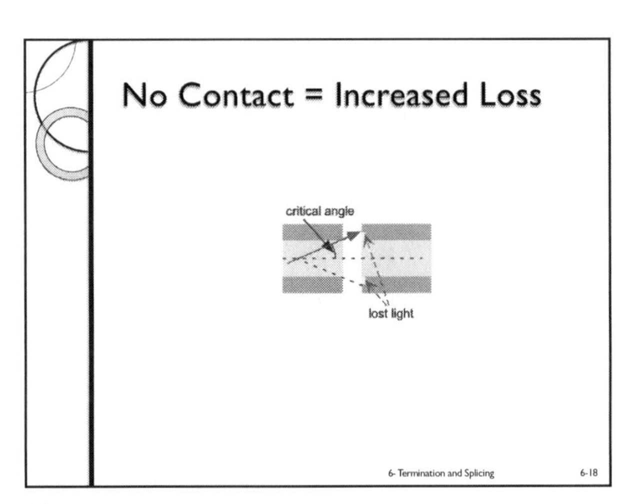

18

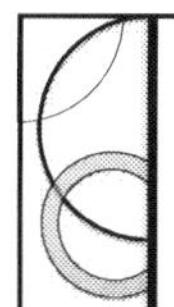

PC, UPC and APC

- Reflectance is *qualitatively* described by
  - PC (physical contact)
  - UPC (ultra physical contact)
  - APC (angled physical contact)
- PC ~ < –40 dB
- UPC ~ < –50 dB
- APC ~ < –55 dB

6- Termination and Splicing 6-19

19

Common Ferrule Diameters

- 2.5 mm
  - ST™-compatible, SC, FC
- 1.25 mm
  - Many small form factor (SFF) and very small form factor (VSFF) connectors have this diameter
  - LC, LX.5, MU
  - Loss of 1.25mm ferrule is lower than that of 2.5 mm ferrule

6- Termination and Splicing 6-20

20

Common Characteristics

- Low to moderate loss
  - Typical loss: 0.15-0.30 dB/pair
- Multiple manufacturers
- Multiple installation methods

6- Termination and Splicing 6-21

21

Basic Connector Information

- Basic Information
- Key Facts
- Function
- Structure
- Colors
- Performance
- Types
- Installation methods

6- Termination and Splicing 6-22

22

Four Functions (5-1)

- Low power loss
  - Deliver sufficient power to receiver
- High retention strength
  - Fix fiber in place
    - To deliver constant power to receiver
- End protection
  - 1 oz of pressure results in 3125 psi on fiber
  - Under this level of pressure, dust on fiber surface can damage surface and increase loss
  - End protection provided by distributing the contact force over area larger than that of fiber to reduce stress on fiber and potential for damage from dust
- Disconnection

6- Termination and Splicing 6-23

23

Connectors

- Basic Information
- Key Facts
- Function
- Structure
- Colors
- Performance
- Types
- Installation methods

6- Termination and Splicing 6-24

24

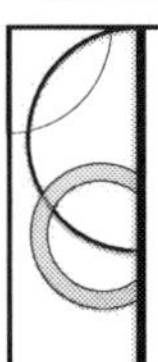

Structure

- Ferrule
- Mating structure
- Simplex or duplex
- Boot
- Dust cap

6- Termination and Splicing 6-25

25

Ferrule

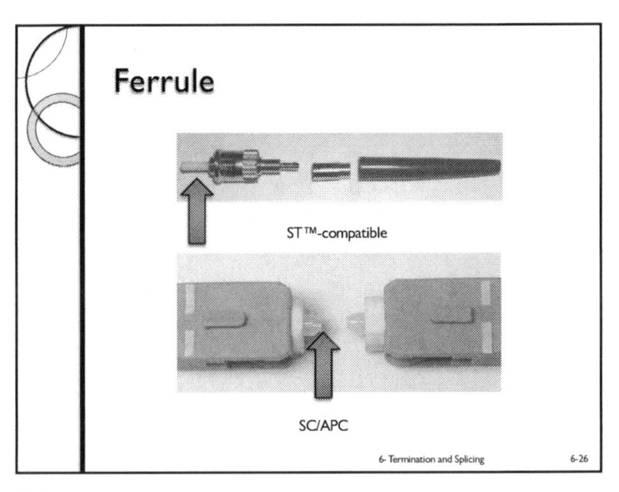

6- Termination and Splicing 6-26

26

Ferrule

- Ferrule provides precise alignment and low power loss
- Low loss results from contacting ferrules
- Connectors can be mated only if the ferrule diameters are the same
  - Example: ST™-compatible and SC both have 2.5mm ferrules
  - Example: SC and LC cannot be mated, as the LC has a 1.25mm ferrule

6- Termination and Splicing 6-27

27

3 Ferrule End Faces

- Radiused
- Angled (APC)
- Flat
  - Not used in current connectors
  - Flat end face produces high reflectance

6- Termination and Splicing 6-28

28

End Face 1: Radius

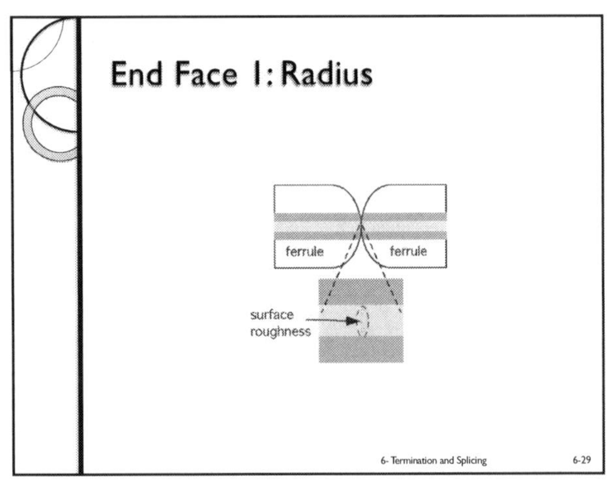

6- Termination and Splicing 6-29

29

Radius End Face

- Eliminates need for perfect perpendicularity on end face
- Radius contact reduces cost
- Known as 'physical contact' (PC)
  - Physical contact (PC) polish reduces both loss and reflectance

6- Termination and Splicing 6-30

30

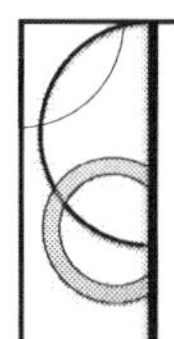

End Face 2: Angle Physical Contact

- Aka 'APC'
- APC is singlemode end face
- End face at 8°angle to perpendicular
- 8° angle provides lowest reflectance possible
  - By reflecting power backwards outside critical angle of singlemode fiber

6- Termination and Splicing 6-31

31

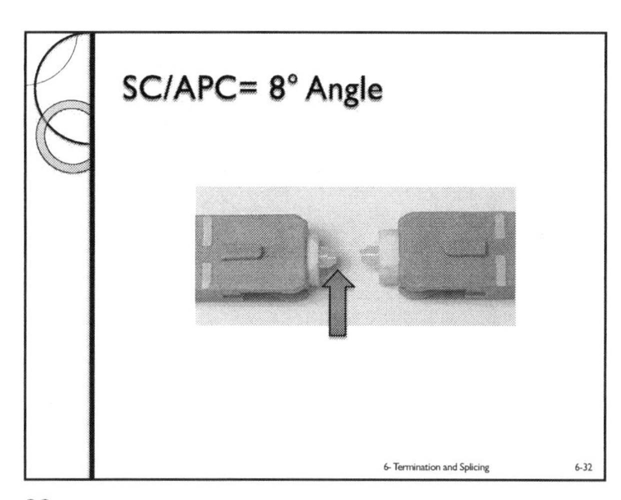

32

Structure Organization

- Ferrule
- Mating structure
- Simplex or duplex
- Boot
- Dust cap

6- Termination and Splicing 6-33

33

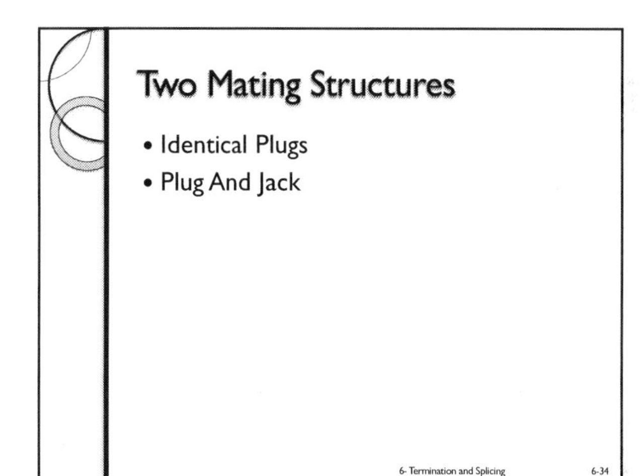

34

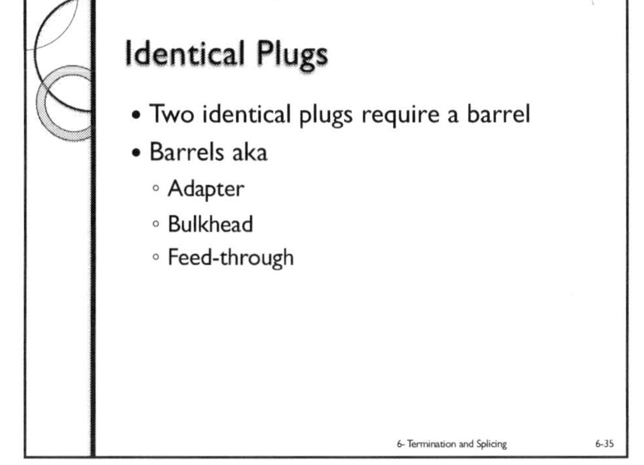

35

Connectors And Barrels

SC

ST™-compatible

6- Termination and Splicing 6-36

36

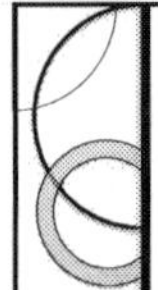

Plug And Jack

- Plug and jack have different structure
  - Plug connectors require the simple 1-lead insertion loss test method
- Jack performs the functions of a plug and barrel in a single structure
  - Plug and jack connectors require 3-lead insertion loss test method (See Testing-10)

6- Termination and Splicing 6-37

37

Plug And Jack

(Courtesy Panduit Corp.)

6- Termination and Splicing 6-38

38

Structure Organization

- Ferrule
- Mating structure
- Simplex or duplex
- Boot
- Dust cap

6- Termination and Splicing 6-39

39

Simplex or Duplex

- Most connectors are simplex
- Current generation connectors are push on, pull off
  - Push on, pull off simplex connectors can be converted to duplex with external clip
- Some very small form factor (VSFF) connectors are intrinsically duplex
  - CS
  - SN
  - MDC

6- Termination and Splicing 6-40

40

LC Simplex

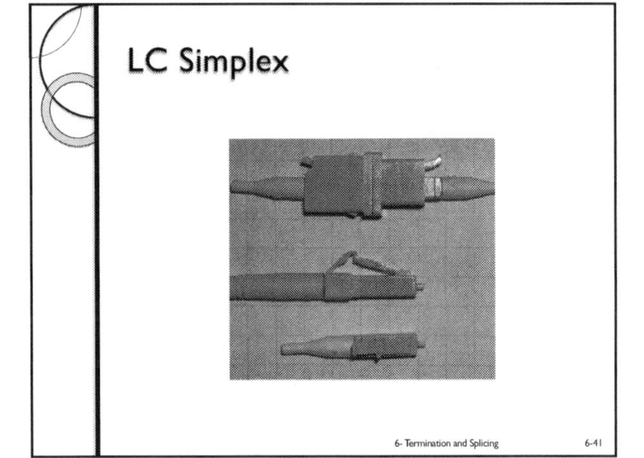

6- Termination and Splicing 6-41

41

LC Duplex With Clip

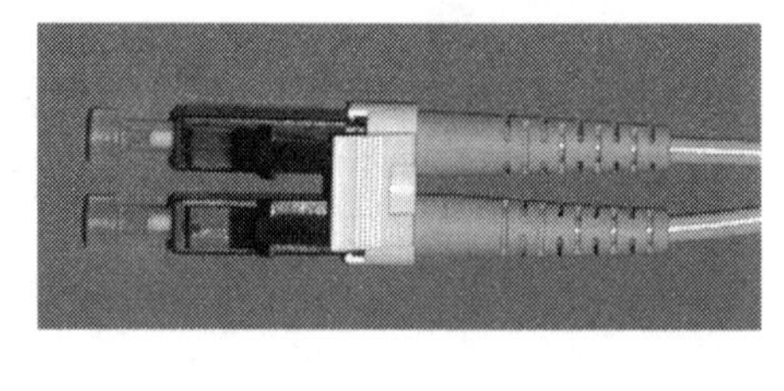

6- Termination and Splicing 6-42

42

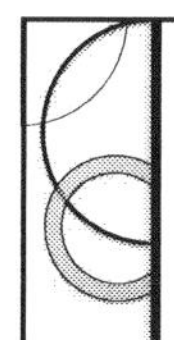

Structure Organization

- Ferrule
- Mating structure
- Simplex or duplex
- Boot
- Dust cap

6- Termination and Splicing 6-43

43

Boot

- The boot controls and limits the bend radius of the fiber as it exits the back shell, providing reliability and low loss

6- Termination and Splicing 6-44

44

Structure Organization

- Ferrule
- Mating structure
- Simplex or duplex
- Boot
- Dust cap

6- Termination and Splicing 6-45

45

Dust Cap

- The cap is the second most important part of the connector (Opinion)
- The cap prevents contamination of the ferrule end
- Such contamination by dust or dirt is reported to result in 85-90% of field problems
- Contamination on contacting ferrules can permanently damage fiber ends
- The rules
  - Keep connectors capped unless plugged into patch panel or transceivers
  - Keep transceivers and patch panel ports plugged when fibers not plugged in

6- Termination and Splicing 6-46

46

Connectors

- Function
- Structure
- Colors
  - On connector body, boot, or barrel
- Performance
- Types
- Installation methods

6- Termination and Splicing 6-47

47

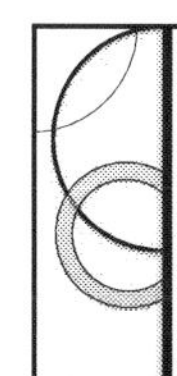

Connector Colors

Color	Type
Green	APC singlemode
Blue	singlemode
Aqua	50 µ LO
Black	50 µ
Beige	62.5 µ
Lime green	OM5

Without color coding, the type of fiber in the connector is unknown

6- Termination and Splicing 6-48

48

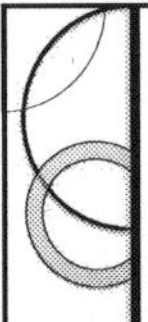

Connectors

- Function
- Structure
- Colors
- Performance
- Types
- Installation methods

6- Termination and Splicing 6-49

49

Optical Performances

- The term 'dB/connection' is used by the Fiber Optic Association
  - A pair creates a 'connection'
    - The term 'dB/pair' is also used
- Maximum insertion loss, dB/connection
- Typical insertion loss, dB/connection
- Repeatability, in dB
- Reflectance, in –dB

6- Termination and Splicing 6-50

50

Maximum Insertion Loss

- Units are "dB/connection"
- The incorrect term, dB/connector, may indicate dB/plug
- Loss is a measurement of power loss is from one fiber to another
  - Must be dB/connection
- Is the maximum loss the installer will experience with correct installation
- This value is a specification
- Installers use this value in certifying networks

6- Termination and Splicing 6-51

51

Consequence of 'dB/Connection'

- Power losses from a transmitter source to a fiber and from a fiber to a receiver are 0 dB (See Ch. 8, Professional Fiber Optic Installation, v10)

6- Termination and Splicing 6-52

52

A Magic Value

- Maximum loss = 0.75 dB/connection
- Note: this value may change in future

6- Termination and Splicing 6-53

53

Optical Performances

- Maximum insertion loss, dB/connection
- Typical insertion loss, dB/connection
- Repeatability, in dB
- Reflectance, in – dB

6- Termination and Splicing 6-54

54

Typical Insertion Loss

- Do not expect to see the maximum value
- The typical insertion loss, in dB/connection, is the value the connector will exhibit when properly installed
- Frequently, the connector will exhibit a value lower than this value
- The designer and the installer use this value in certifying networks
- You want to know this value, but it is not a specification (See: Section 8, Acceptance Values Certification)

6- Termination and Splicing 6-55

55

Three Typical Loss Values

- Pre-polished connector with 2.5 mm ferrule
  - ~ 0.40 dB/connection
- Pre-polished connector with 1.25 mm ferrule
  - ~ 0.20 dB/connection
- Splice on connector with 1.25 mm ferrule
  - ~ 0.20 dB/connection

6- Termination and Splicing 6-56

56

Optical Performance Concerns

- Maximum insertion loss, dB/connection
- Typical insertion loss, dB/pair
- Repeatability, in dB
- Reflectance, in – dB

6- Termination and Splicing 6-57

57

Repeatability

- Repeatability is the change in connector loss between successive connections
  - Unit is dB
- Connectors have repeatability performance
  - Splices are not disconnected and do not have a repeatability characteristic
- If range not measured, repeatability is used to calculate range

6- Termination and Splicing 6-58

58

Range

- We do not expect exact same loss from measurement to measurement
- Range is the maximum increase in loss expected between successive connections
- Range is used to interpret increase in loss
  - If an increase in loss is less than the range, the increase indicates normal behavior
  - If an increase in loss is greater than the range, the increase indicates degradation and a need for troubleshooting

6- Termination and Splicing 6-59

59

Optical Performance Concerns

- Maximum insertion loss, dB/connection
- Typical insertion loss, dB/connection
- Repeatability, in dB
- Reflectance, in – dB

6- Termination and Splicing 6-60

60

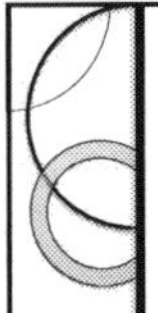

Reflectance = Fresnel Reflection

- Reflectance occurs at a glass air interface
- This interface, or any interface at which there is a change in speed of light, or index of refraction, can produce a reflection, called a "Fresnel reflection"
- This reflection occurs in fiber optic connectors because the end faces of mated connectors have surface roughness, which creates microscopic air gaps
- Reflectance is as important as loss for singlemode connectors but rarely important for multimode connectors

6- Termination and Splicing 6-61

61

Microscopic Air Gaps Result in Reflectance Due to Less than Perfect Polishing

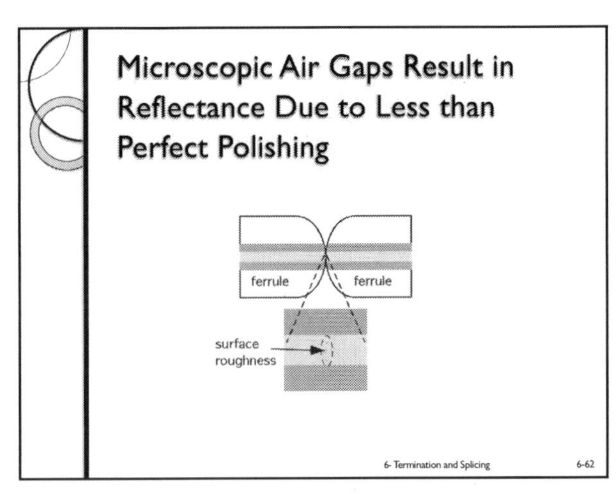

6- Termination and Splicing 6-62

62

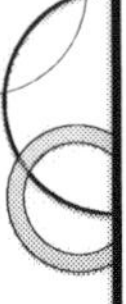

Reflectance 1

- A measurement of the relative optical power reflected backwards from a connector
- Reflected power can travel back to the transmitter and be reflected from the source back into the fiber
- If this reflected power reaches the receiver with a power level above its sensitivity, the receiver will convert this optical power to a digital 'one'

6- Termination and Splicing 6-63

63

Reflectance 2

- Reflectance influences the accuracy with which a fiber optic link will transmit
- Minimizing reflectance results in maximizing signal accuracy
- If the time interval in which the reflected power arrived was a digital 'zero' at the transmitter, and if the reflected power level is above the sensitivity of the receiver, the output signal will differ from the input signal
- The objective of a reflectance requirement is to limit the reflected power at the receiver to a value less than the sensitivity of that receiver

6- Termination and Splicing 6-64

64

In Other Words

- Excessive reflectance creates high bit error rates, long transmission times, or inability of link to transmit

6- Termination and Splicing 6-65

65

Reflectance Definition

- Definition
  - 10 log (reflected power/incident power)
- Reflectance is stated in units of negative dB
- Reflectance requirements are from -20 dB to -65 dB
  - -20 dB for multimode connectors at ≥ 1 Gbps
  - -65 dB for APC connectors (singlemode)

6- Termination and Splicing 6-66

66

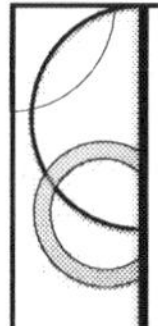

PC, UPC and APC

- Reflectance is *qualitatively* described by
  - PC (physical contact)
  - UPC (ultra physical contact)
  - APC (angled physical contact)
- PC ~ < –40 dB
  - PC finish of radius connector end reduces both loss and reflectance
- UPC ~ < –50 dB
- APC ~ < –55 dB

6- Termination and Splicing 6-67

67

Typical Reflectance Values

- The Fresnel reflection of fiber to air
  - -14 dB to -18 dB (from a non-contact connector)
- From hand polishing singlemode connector
  - < -50 dB
- From factory polished singlemode connector
  - < -55 dB
- APC
  - < -64 dB

6- Termination and Splicing 6-68

68

Recommendation

- Reflectance is of concern on singlemode connectors
- Whenever reflectance is of concern, fusion splice factory-polished singlemode pigtails
  - Instead of field polishing connectors
- Exceptions
  - Splice on connectors
  - Cleave and crimp connectors
  - Reason: fiber end is factory polished

6- Termination and Splicing 6-69

69

Connectors

- Function
- Structure
- Colors
- Performance
- Types
- Installation methods

6- Termination and Splicing 6-70

70

Four Connector Groups

- Small Form Factor (SFF)
- Very Small Form Factor (VSFF)
- Commonly used types
- Legacy types

6- Termination and Splicing 6-71

71

Four Connector Groups

- Small Form Factor (SFF)
- Very Small Form Factor (VSFF)
- Commonly used types
- Legacy types

6- Termination and Splicing 6-72

72

SFF Advantage

- Increased connector count in reduced patch panel space
- Small form factor (SFF) connectors are dominant

6- Termination and Splicing 6-73

73

Obvious Size Reduction

SC/APC

MU

LC

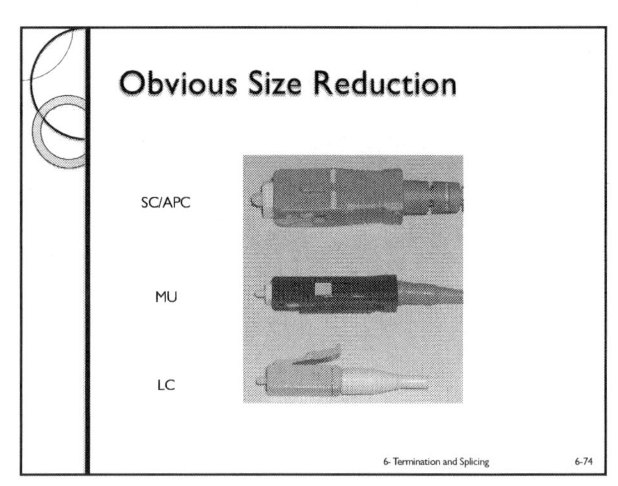

6- Termination and Splicing 6-74

74

Common SFF Connectors

- LC
- LX.5
- MU
- Volition
- MTP/MTO

6- Termination and Splicing 6-75

75

LC Simplex

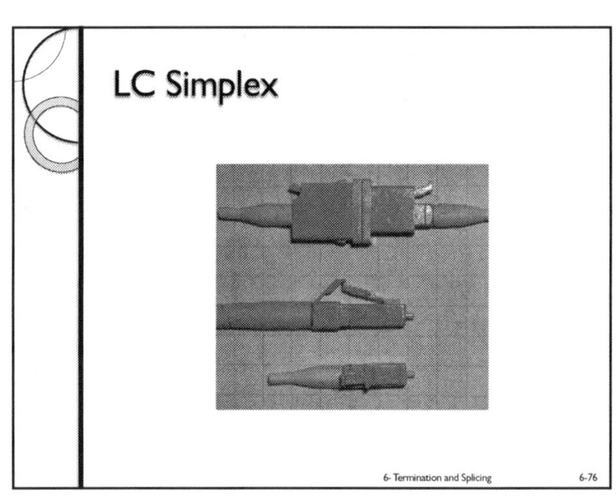

6- Termination and Splicing 6-76

76

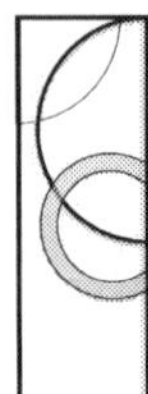

LC Duplex With Clip

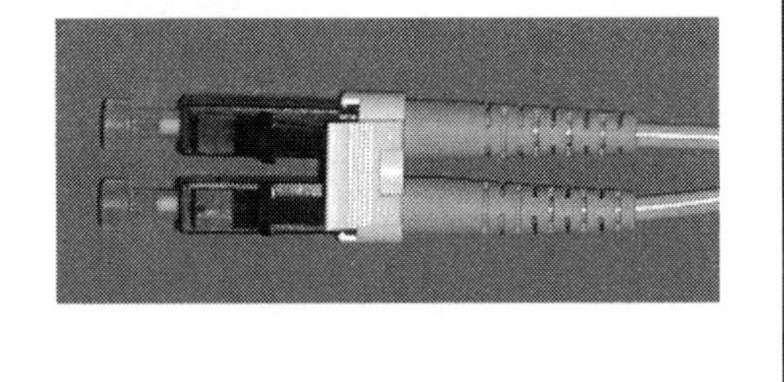

6- Termination and Splicing 6-77

77

LC Characteristics

- Low loss: typical 0.15-0.20 dB/pair
- Pull proof and wiggle proof design
- 1.25 mm ferrule provides reduced loss
- Preferred interface for 10 Gbps optoelectronics (author's opinion)

6- Termination and Splicing 6-78

78

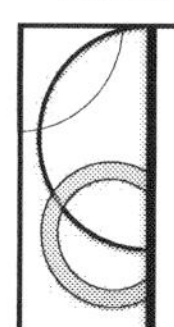

Recommendation

- Choose LC
- Reasons
  - Low loss
  - 22 years of use has revealed no hidden problems
  - Available from multiple manufacturers
  - Available in multiple installation methods

6- Termination and Splicing 6-79

79

LX.5

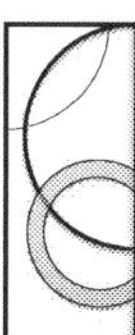

6- Termination and Splicing 6-80

80

LX.5 Background

- Developed by ADC Telecommunications, now owned by CommScope
- Similar to LC
  - 1.25 mm ferrule
  - Available in radius and APC versions

6- Termination and Splicing 6-81

81

LX.5 Characteristics

- Low loss: typical 0.15-0.20 dB/connection
- Pull proof
- Wiggle proof design
- 1.25 mm ferrule with low loss
- Unique features
  - Built in dust covers in
    - Connector
    - Barrel
  - Act as dust covers and as eye safety shutters

6- Termination and Splicing 6-82

82

LX.5 Built In Ferrule Cover

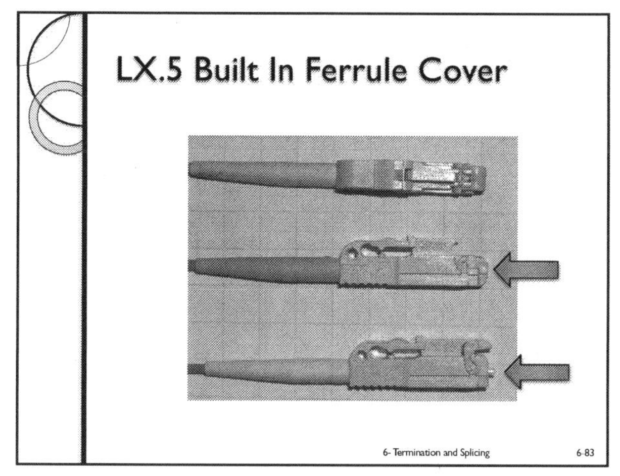

6- Termination and Splicing 6-83

83

LX.5 Built In Barrel Shutter

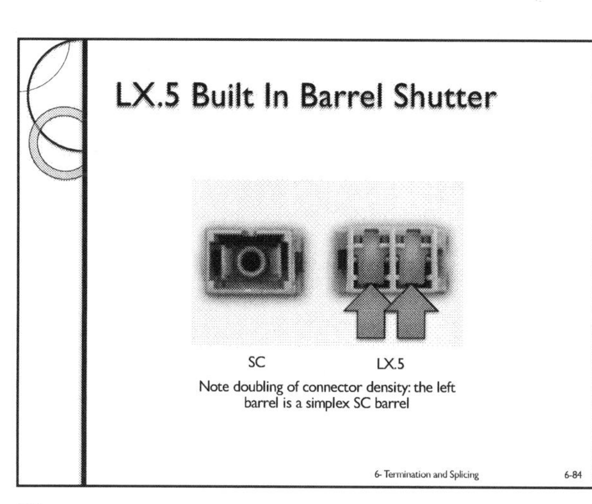

Note doubling of connector density: the left barrel is a simplex SC barrel

6- Termination and Splicing 6-84

84

MU

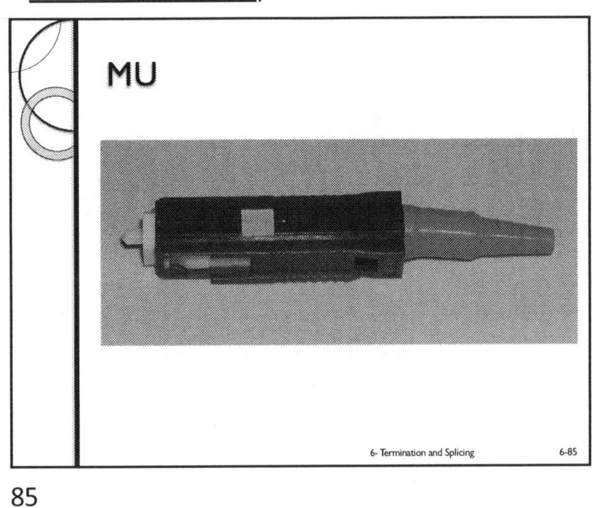

6- Termination and Splicing 6-85

85

MU Characteristics

- The appearance of an SC but with all dimensions reduced by 50 %
- 1.25mm ferrule with low loss
- A simplex connector
  - Unique feature: can create a duplex, triplex or quadraplex connector
- Pull proof, wiggle proof design

6- Termination and Splicing 6-86

86

Volition

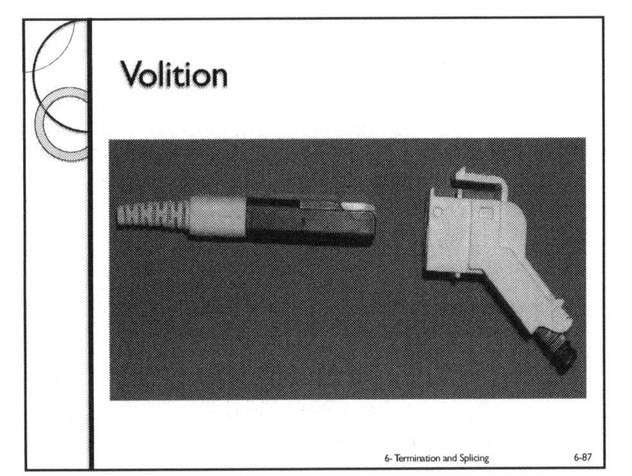

6- Termination and Splicing 6-87

87

Volition Background

- Plug and jack system
- A duplex SFF connector
- Plug and jack design
- Estimated number at George Washington University: 66,000-88,000
- Sole source: 3M

6- Termination and Splicing 6-88

88

Volition Characteristics

- Low loss: typical 0.3 dB/connection(field installed)
- Pull proof
- Wiggle proof design
- Unique design feature: 'V' grooves instead of ferrules
- Fibers in plug float in space
- Fibers in jack in V grooves

6- Termination and Splicing 6-89

89

Volition™ Plug

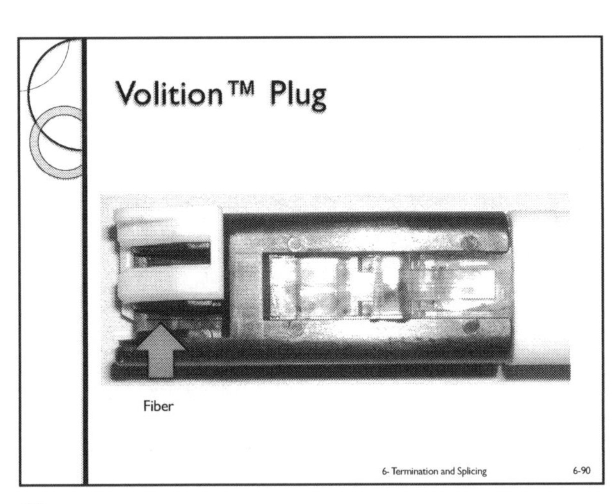

Fiber

6- Termination and Splicing 6-90

90

Volition™ Jack

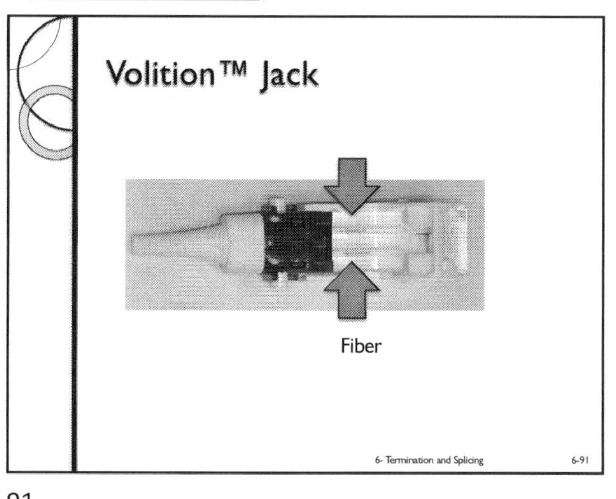

6- Termination and Splicing 6-91

91

MTP/MPO

- By some, the MTP/MPO is not considered an SFF connector
- However, with 12-72 fibers in volume of that of a single fiber SC, MTP®/ MPO qualifies as a SFF connector (opinion)

6- Termination and Splicing 6-92

92

MTP– SC Size Comparison

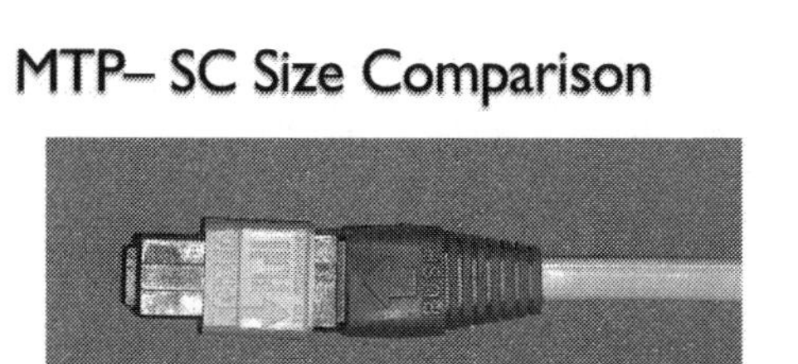

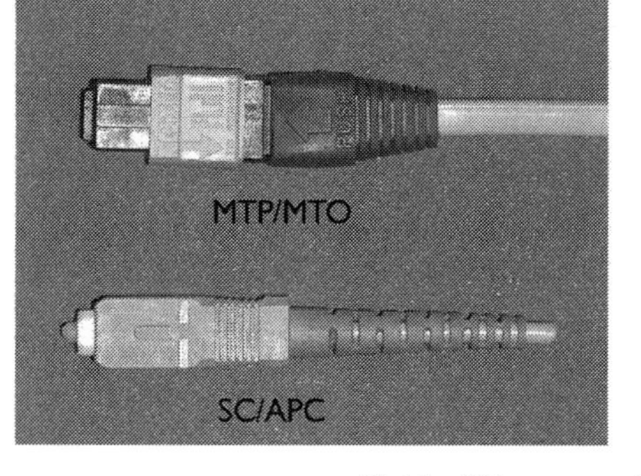

6- Termination and Splicing 6-93

93

MTP®/MPO Connectors

- MPO: Multiple-Fiber Push-On/Pull-Off
- Used in pre-terminated systems
  - Indoors in data centers
  - Outdoors in FTTH networks
- Considered 'ribbon' connector
  - 8, 16, 12, 24, 72 fiber versions available
  - Can be used with 12 fiber/tube, loose tube cables

  - MTP® is a registered trademark of US Conec Ltd.

6- Termination and Splicing 6-94

94

MTP/MPO: 12 Fibers

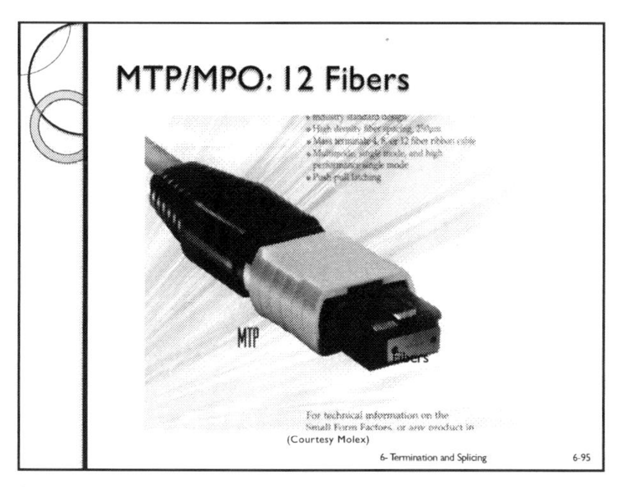

(Courtesy Molex)

6- Termination and Splicing 6-95

95

MTP®/MPO

- MTP®/MPO used
  - As patch cables with MTP® /MPO connectors on both ends
  - As break out cables, with individual single fiber connectors on opposite end (e.g., LCs)

6- Termination and Splicing 6-96

96

MTP®/MPO-LC Break Out Cable

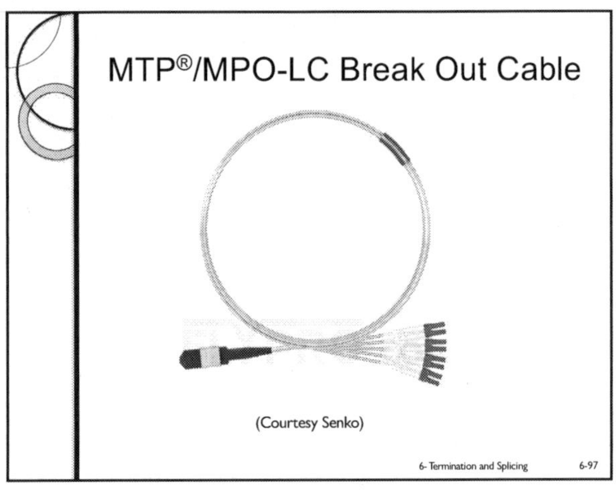

(Courtesy Senko)

6- Termination and Splicing 6-97

97

MTP/MPO Characteristics

- For 12 fiber versions
  - Typical loss: ≤0.5 dB/connection, multimode
  - Typical loss: ≤0.75 dB/connection, singlemode
- Fibers protrude above ferrule end
  - Connector requires factory installation
  - Cleaning requires increased care
- Available in radius and APC versions

6- Termination and Splicing 6-98

98

Advantages

- Reduced installation cost
  - Data centers
  - Riser networks
- Reduced network cost
  - Reduced patch panel cost and space

6- Termination and Splicing 6-99

99

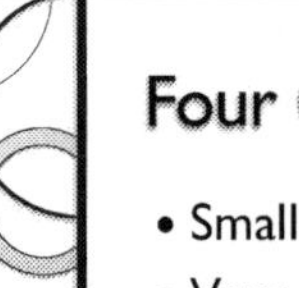

Four Connector Groups

- Small form factor (SFF)
- Very Small Form Factor (VSFF)
- Commonly used types
- Legacy types

6- Termination and Splicing 6-100

100

VSFF Connectors

- Used in data centers to achieve very high connector count in small patch panel area
- Very small form factor connectors (VSFF) increase the fiber count in reduced space
- Example:
  - In 1 RU (RU, rack unit= 1.75")
  - SC: 48
  - LC: up to 144
  - CS: 336

6- Termination and Splicing 6-101

101

VSFF Size Reduction

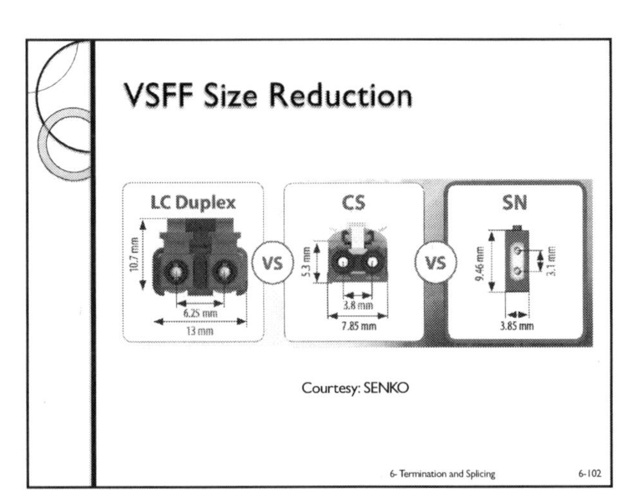

Courtesy: SENKO

6- Termination and Splicing 6-102

102

Four Connector Groups

- Small form factor (SFF)
- Very Small Form Factor (VSFF)
- Commonly used types
- Legacy types

6- Termination and Splicing 6-103

103

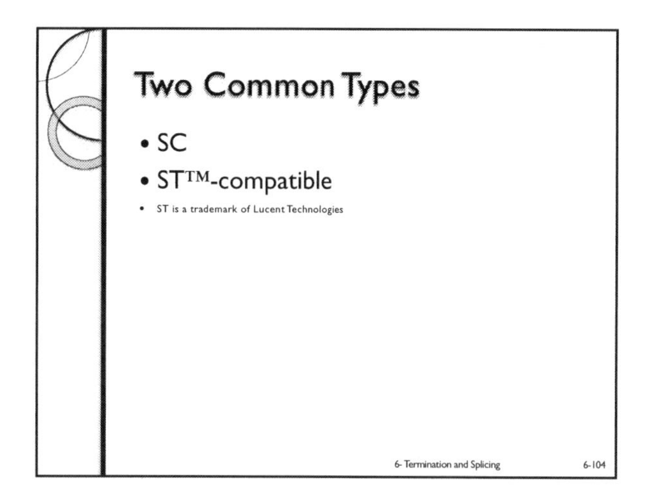

104

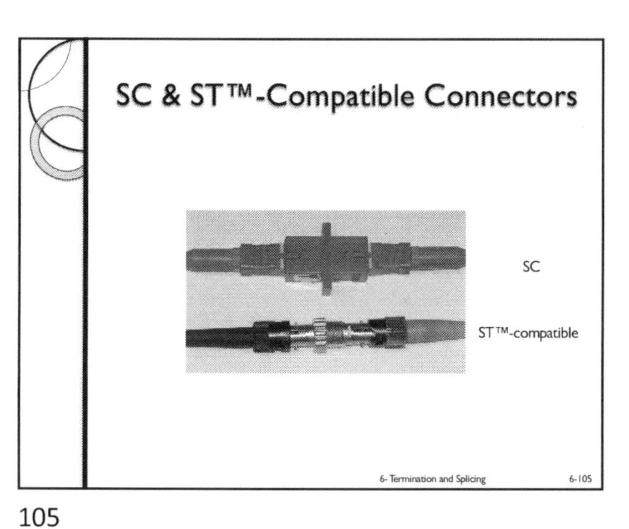

105

SC & ST™ Comparisons

- Both: contact, keyed, 2.5mm ferrules, low typical loss (~0.3 dB/connection)
- Latching
  - SC- push on, pull off; ≤ 48/1U enclosure; pull and wiggle proof
  - ST™-twist on and off; 12/1U enclosure; not pull or wiggle proof

6- Termination and Splicing 6-106

106

Four Connector Groups

- Small form factor (SFF)
- Very Small Form Factor (VSFF)
- Commonly used types
- Legacy types

6- Termination and Splicing 6-107

107

Legacy Connector Characteristics

- Were used in the past but are no longer used for new installations
- See Ch. 5, Professional Fiber Optic Installation, v10 for information

6- Termination and Splicing 6-108

108

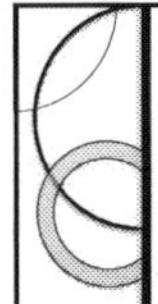

Connectors

- Function
- Structure
- Colors
- Performance
- Types
- Installation methods

6- Termination and Splicing 6-109

109

Connector Installation Methods

- Epoxy
  - The first method developed
  - Used with connectors/termination installed on factory made patch cords
- Fusion spliced pigtails
  - Most common method for installing singlemode connectors is with fusion splicing of pigtails
  - Also splice on (SOC) are used
- 'Cleave-and-Crimp' Installation
- Hot Melt™
- Quick Cure Adhesive
  - Not commonly used

6- Termination and Splicing 6-110

110

Epoxy Installation (5-13)

- In 1970's, first method developed
- High reliability: epoxy withstands wide range of environment conditions
- High process yields
- Used in many cable assembly facilities

6- Termination and Splicing 6-111

111

Fusion Spliced Pigtails

- The connector installation method with the lowest total installed cost does not require connector installation
  - Fusion splice factory-installed pigtails,
  - Almost all singlemode connectors are installed as fused pigtails
  - A pigtail is a short length of jacketed cable or 900 μ fiber with a connector on one end
  - Cut a patch cord in half to create two pigtails
- Quality of cleaved fiber end determines loss
- Reduction in labor cost pays for fusion splicer in ~< 1000 connectors

6- Termination and Splicing 6-112

112

Pigtail Benefit

1. 100 % yield
2. High reliability
3. Low reflectance

6- Termination and Splicing 6-113

113

Common Feature

- Most methods require polishing
- Polishing is the step at which
  - 90-95 % of problems occur
  - Most of the time is spent
- If we could eliminate polishing, we could
  - Eliminate most problems
  - Reduce installation time and labor cost

6- Termination and Splicing 6-114

114

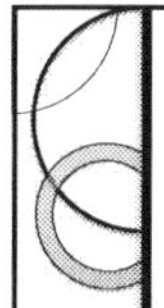

Splice On Connectors (SOCs)

- Are pre-polished connectors
  - That is, they do not require field polishing
- Installation requires making splice
  - Quality of cleave (low cleave angle) determines loss of connector
- Two types
  - Cleave and crimp, a mechanical splice within the connector (Example: Unicam®)
  - Fuse on, a fusion spliced on connector (aka SOC)

6- Termination and Splicing 6-115

115

'Cleave-and-Crimp' (5-16)

- Connectors have a pre-installed fiber
- Outside fiber end is pre-polished
- Quality of cleave (low cleave angle) determines loss of connector
- Index matching gel is on inside end of fiber
- No polishing required!
- This is a mechanical splice in the connector back shell

6- Termination and Splicing 6-116

116

'Cleave-and-Crimp' Description

- Prepare cable end
- Clean fiber
- Cleave fiber
- Insert fiber
- Crimp connector to fiber and buffer tube
- Done

6- Termination and Splicing 6-117

117

'Cleave-and-Crimp' Advantages

- Fast installation
  - Web pages claim: 30/hour, 40/hour and 60/hour
  - I believe 30, want to see 40, and want to video tape 60!
- Acceptable yield
  - Experienced installer: 90-95 %
- Key advantage: reduced training time and cost
- Reduced set up and clean up time
  - Critical for few connectors per location
- Installation can be done in almost any environment

6- Termination and Splicing 6-118

118

'Cleave-and-Crimp' Disadvantages 1

- Typical loss can be higher than connectors that require polishing
  - 0.40 dB/connection for 2.5mm ferrules
  - 0.25 dB/connection for 1.25mm ferrules
  - Vs. 0.1-0.3 dB /pair for polish connectors
- Troubleshooting difficult
  - Usually, you will not know cause of high loss

6- Termination and Splicing 6-119

119

'Cleave-and-Crimp' Disadvantages 2

- Connector cost is high
  - $11-13 for multimode SC
  - Vs. $5.50 for multimode Hot Melt SC
- Total installed cost is high
  - Price premium is not offset by reduced labor cost unless total loaded labor rate is very high
- Process yield depends on
  - Installer

6- Termination and Splicing 6-120

120

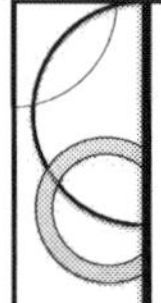

Splice On Connectors (SOCs)

- Connector has pre-installed fiber protruding beyond back shell
- Requires
  - Use of compatible splicing machine
  - Removable fiber and SOC holders
- Yield close to 100 %
- Use when enclosure space is insufficient for splice trays

6- Termination and Splicing 6-121

121

Splice On Connectors 2

- Also known as Fuse On
- Enables change of connector type
- Good performance and yield in field
  - Typical loss 0.2-0.3 dB/connection
- Connectors cost ~ $10-20
- Shipboard applications favor SOC use

6- Termination and Splicing 6-122

122

Hot Melt™ Adhesive (5-14)

- Developed by 3M to
  - Improve convenience
  - Reduce time required to install connectors
  - Reduce labor cost
  - 100 % yield possible
- Sole source
- Connectors preloaded with adhesive

6- Termination and Splicing 6-123

123

Quick Cure Adhesive (5-15)

- Two part anaerobic-like adhesive
- Developed in early 1990's by
  - Lucent Technologies
  - Automatic Tool And Connector Co. (now Suttle Industries Inc.)

6- Termination and Splicing 6-124

124

Hands-On Connector Installation

- Do not proceed until you hear the magic word: Proceed
- Multimode mechanical splice-on connector (Unicam®)
  - Not included in Advanced Splicing delivery
- Singlemode fusion splice-on connector (SOC)
- Singlemode fusion spliced pigtail

6- Termination and Splicing 6-125

125

Termination Part 2

SPLICES

6- Termination and Splicing 126

126

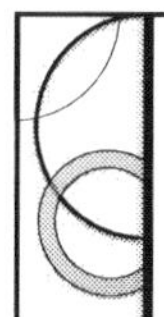

Essential Splice Information

- Definition
- Function
- Locations
- Process
- Types
- Splice Hardware
- Performance

6- Termination and Splicing 6-127

127

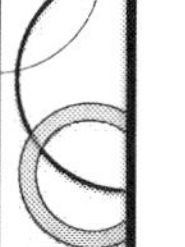

Definition

- A splice is preferred when there is no need for disconnection or rerouting at the location of the connection
- Splicing is the process of making a permanent connection between fibers
  - Some mechanical splices allow semi-permanent connections
- Splicing can be
  - Single fiber splicing
  - Ribbon fiber splicing (aka, mass splicing)
    - Ribbon splicing reduces splicing cost

6- Termination and Splicing 6-128

128

Essential Splice Information

- Definition
- Functions
- Locations
- Process
- Types
- Hardware
- Performance
- Language

6- Termination and Splicing 6-129

129

Two Splice Functions

- Low loss
- High strength

6- Termination and Splicing 6-130

130

Essential Splice Information

- Definition
- Functions
- Locations
- Process
- Types
- Hardware
- Performance
- Language

6- Termination and Splicing 6-131

131

Two Locations And Reasons

- Mid Span
  - The cable cannot be obtained in a single length
  - The cable cannot be installed in a single length
  - The cable has experienced backhoe fade, post hole driller fade, rodent fade, shark fade or any other source of damage than creates link failure by broken fibers
- Pigtail
  - Pigtails are short lengths of cables or tight buffer tubes with a factory-connector installed on one end
  - Installers perform pigtail splicing to reduce installation cost or installation time
  - The majority of singlemode connectors are spliced pigtails, fuse on, or splice on connectors

6- Termination and Splicing 6-132

132

Splice Locations

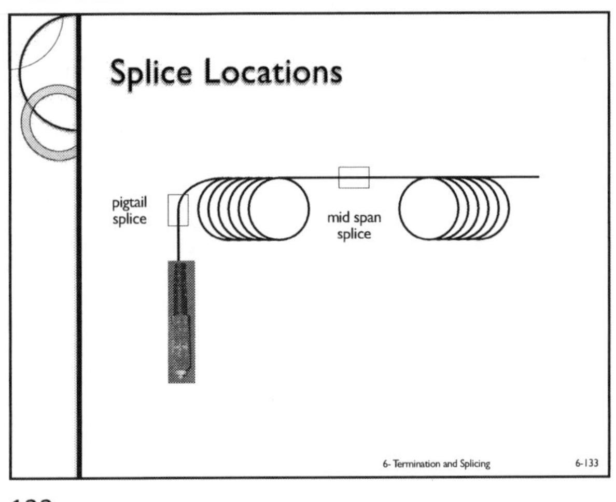

6- Termination and Splicing 6-133

133

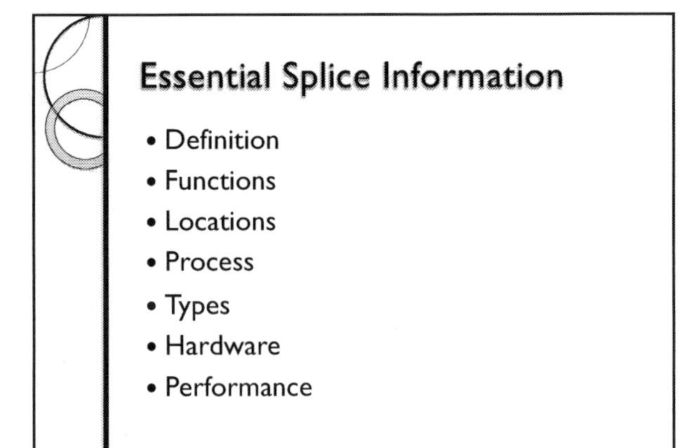

Essential Splice Information

- Definition
- Functions
- Locations
- Process
- Types
- Hardware
- Performance

6- Termination and Splicing 6-134

134

Splicing Process

- Prepare enclosure
- Prepare cable end
- Place fiber and buffer tubes in enclosure
- Make the splice
- Place splice and fiber into the tray
- Close and seal the enclosure

6- Termination and Splicing 6-135

135

Essential Splice Information

- Definition
- Functions
- Locations
- Process
- Types
- Hardware
- Performance

6- Termination and Splicing 6-136

136

Two Splice Types

- Fusion
- Mechanical

6- Termination and Splicing 6-137

137

Tools Required

- Almost the same tool kit for both methods
  - Base kit~ $800
  - Main cost is a high precision cleaver (≤$600)
  - Note: cleavers @ ≤ $200 provide acceptable results
- Fusion splicing requires fusion splicer
  - $3000-$9000 (7/15/23)
- Mechanical splicing may require an inexpensive tool (<$100)

6- Termination and Splicing 6-138

138

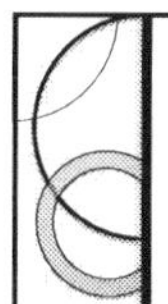

Fusion Splicing

- Is the process of fusing, or welding together, of two fibers
  - Aka glass welding
- Labor cost savings can recover cost of fusion splicer in ≤1000 connectors (author's estimate)

6- Termination and Splicing 6-139

139

Two Splicer Functions

- The precise alignment of the fibers to each other prior to splicing
  - Precise alignment means sub micron precision
  - Alignment results in low power loss
- Low loss results from
  - Low angle cleaves
  - Precise control of the splicing operation

6- Termination and Splicing 6-140

140

Two Alignment Methods

- Passive
  - Cost: $2,200-$3,000
- Active
  - Cost : $3,000-$9,000
  - Ribbon splicer cost: $12,000-$20,000

6- Termination and Splicing 6-141

141

Passive Alignment

- Based on a precision 'V' groove
- Assumptions
  - Fiber diameters and the
  - Core-cladding concentricity are precise enough to achieve low power loss
- These assumptions are valid for both the fiber made in North America and much (but not all) of the fiber made overseas

6- Termination and Splicing 6-142

142

V-Groove

6- Termination and Splicing 6-143

143

Active Alignment

- Active means fibers are moved to provide low loss
- Movement compensates for fiber differences
  - Core, MFD, cladding diameters
  - Core offsets
  - Cladding non-circularity

6- Termination and Splicing 6-144

144

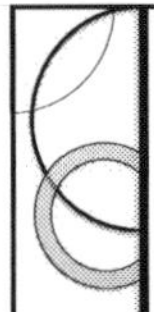

Two Active Alignment Methods

- For singlemode fiber
  - Profile alignment (aka PAL and PAS)
    - Dominant method
    - From Fujikura, Sumitomo, Fitel
  - Local injection and detection (LID)
    - From Corning Cable Systems (Siemens product)
- Both methods provide same
  - Low loss
  - High strength

6- Termination and Splicing 6-145

145

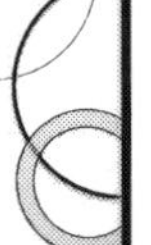

PAS

- Collimated light is passed through both fibers
- Collimated light reveals core-cladding boundary
- Microscope lens behind fibers collects image
- Image is digitized
- Core cladding boundary identified
- Fibers moved in X and Y axes to minimize misalignment of core cross section

6- Termination and Splicing 6-146

146

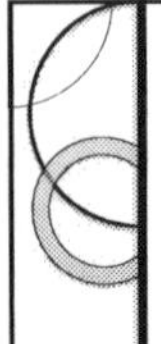

PAS Alignment Mechanism

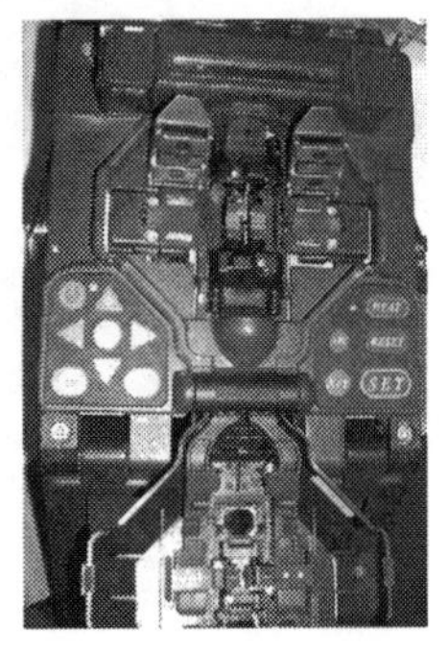

6- Termination and Splicing 6-147

147

PAS Final Screen

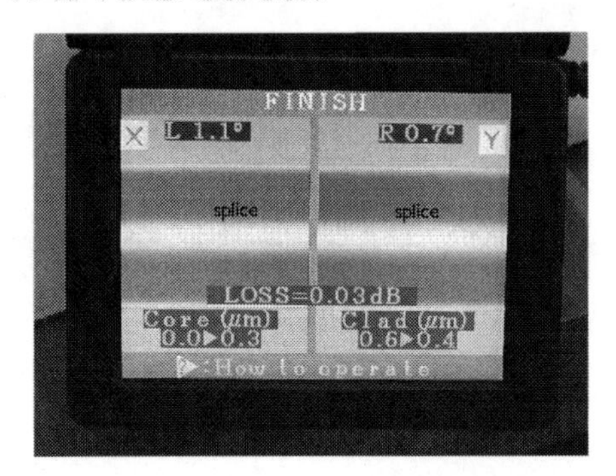

6- Termination and Splicing 6-148

148

LID

- Light is launched into one fiber and tapped from the second fiber
- Fibers are moved to provide maximum power transfer between two fibers

6- Termination and Splicing 6-149

149

Multimode Alignment

- Is cladding alignment
- Is sufficient for multimode cores
  - Multimode core area is ~37 times larger than that of singlemode core
  - Multimode fiber does not require compensation for
    - Core offset
    - Cladding non-circularity

6- Termination and Splicing 6-150

150

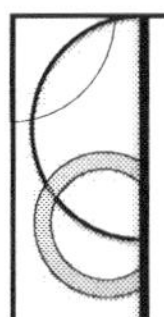

Loss Estimation

- Splicer calculates the splice loss from dimensions of fiber at the splice
- Estimation used to decide whether to put splice in tray
  - Estimate is calculation, not measurement
  - OTDR loss is a measurement
- Accurate OTDR loss is average measurements in opposite directions

6- Termination and Splicing 6-151

151

Fusion Splicing Used In

- Most initial installations
- In restoration

6- Termination and Splicing 6-152

152

4 Fusion Splicing Advantages

- Low power loss
  - Often, fusion splices result in 0 dB loss
- Low to no reflectance
  - Theory says 'low' reflectance
  - No reflectance is the rule
- High strength
- Low cost per splice
  - Only unique consumable is fusion splice cover @ ~ $0.50/splice

6- Termination and Splicing 6-153

153

Two Fusion Splicing Disadvantages

- Fusion splicer can be expensive
- If the number of splices is low, fusion splicing may be cost prohibitive, even with rented splicer
  - $900 rental charge for 100 splices= $9/splice
  - A mechanical splice may be less expensive

6- Termination and Splicing 6-154

154

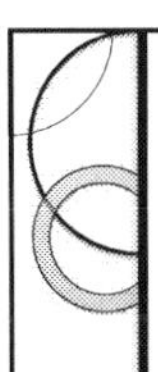

One Additional Disadvantage: No Fusion Splicing In Manhole

-manholes collect methane!

6- Termination and Splicing 6-155

155

Mechanical Splicing

- Is the process of placing two cleaved ends in a mechanical splice
  - Low angle cleaves result in low loss
- Splice provides cladding alignment and retention, or strength
- The mechanical splice provides the function of a cover

6- Termination and Splicing 6-156

156

Mechanical Splice Types

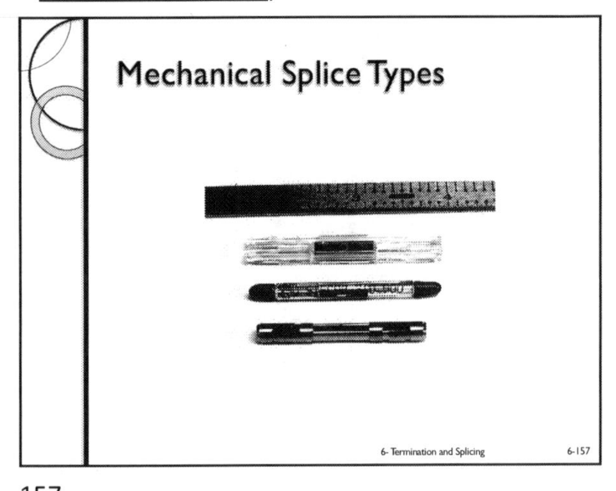

157

Alignment Methods

- Some splices have a precision capillary tube for alignment
- Other splices use a precision etched silicon substrate
- The 3M FibrLok2® has a precision 'V' groove

158

Gripping Details

- Most mechanical splices grip by compression on the fiber
  - Some grip with drill chuck type mechanism
  - Some grip be closing spring loaded mechanism
- Some use adhesive or epoxy

159

RI Matching Gel

- All mechanical splices have RI matching gel at their centers
- Gel fills any air gap that results from the end faces of being less than perfectly perpendicular (which they are)
- Such filling reduces loss
- In singlemode fibers, the gel can eliminate all reflectance
  - When 3 RIs are the same

160

Advantages

- Organizations that have many splice teams
  - Advantage: reduced capital equipment cost
- Small number of splices
  - Advantage: reduced cost per splice

161

Recommendation (Opinion)

- Use FibrLokII (p/n 2529) from 3M™
- Consistent low loss
  - Typical loss, singlemode or multimode, of 0.05 dB
  - Even with mismatched fibers
  - 2529 can be used on both singlemode and multimode fibers!

162

Advantages

- Low capital equipment cost
- Many are re-enterable or reusable
- Some can be tuned to low loss with VFL

6- Termination and Splicing 6-163

163

Mechanical Splice Disadvantage

- Major disadvantage is high cost in large installations
  - At a price of $13-$30, a 1000 splice installation will cost $13,000-$30,000
- At $5000-$9,000, a fusion splicer becomes attractive alternative

6- Termination and Splicing 6-164

164

Essential Splice Information

1. Definition
2. Locations
3. Function
4. Steps
5. Types
6. Splice Hardware Components
7. Performance

6- Termination and Splicing 6-165

165

Three Primary Splice Components

- Splice and cover
- Splice holder
- Splice tray
- Splice enclosure

6- Termination and Splicing 6-166

166

Splice Cover

- The installer places a splice cover over the fusion splice
- The cover isolates the splice from the environment, supports, and protects the splice
- Splice covers can be heat shrinkable or adhesive

6- Termination and Splicing 6-167

167

Shrinkable Splice Covers & Holder

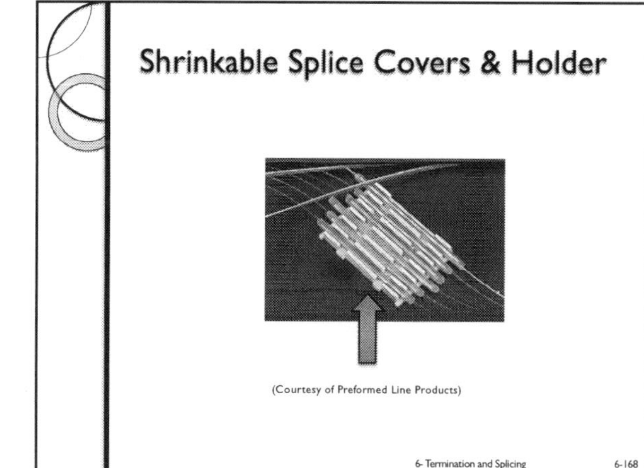

(Courtesy of Preformed Line Products)

6- Termination and Splicing 6-168

168

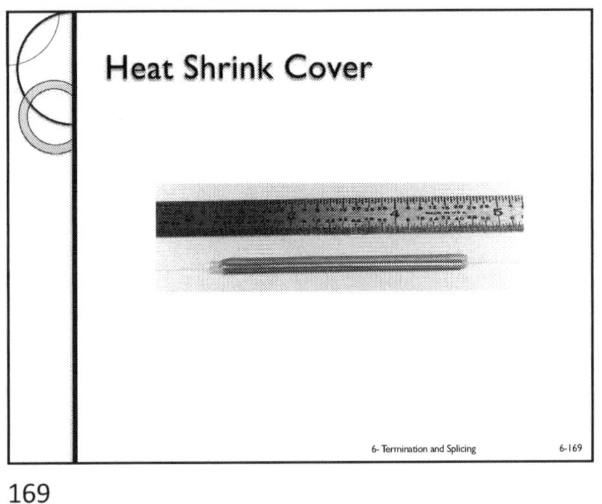

Heat Shrink Cover

6- Termination and Splicing 6-169

169

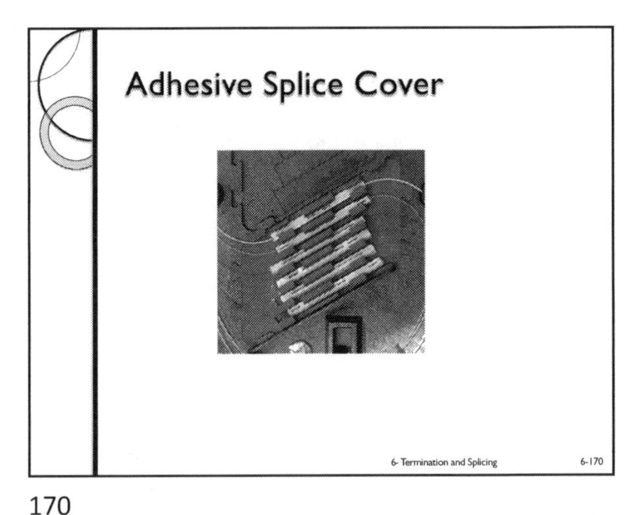

Adhesive Splice Cover

6- Termination and Splicing 6-170

170

Splice Tray

- The tray protects the splice and houses excess fiber
- Excess fiber required to place fiber in splicer

6- Termination and Splicing 6-171

171

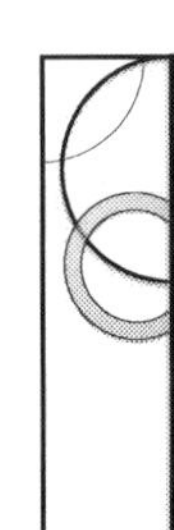

Half Length Splice Tray

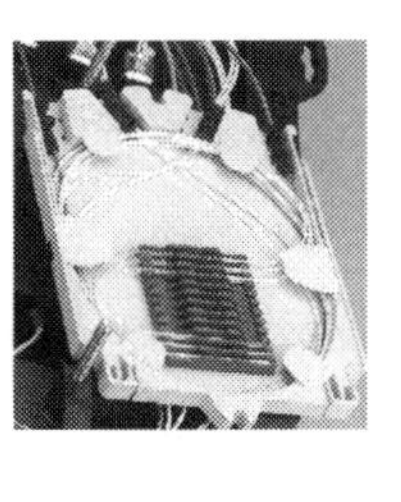

(Courtesy of Preformed Line Products)

6- Termination and Splicing 6-172

172

Full Length Splice Tray

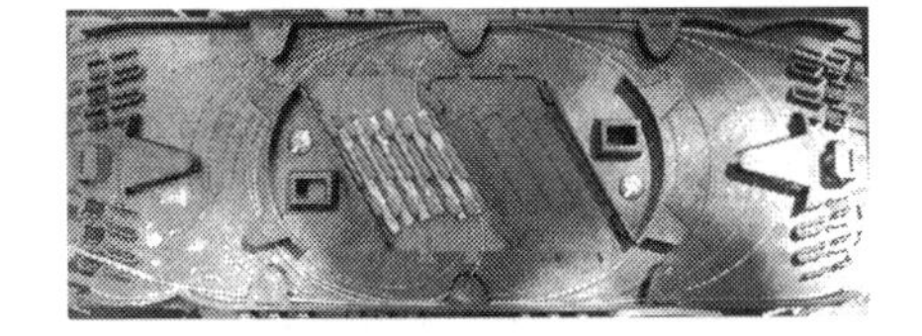

6- Termination and Splicing 6-173

173

Enclosure

- The enclosure houses and protects the splice trays and excess buffer tube length
- Excess buffer tube allows the splice tray to be outside of the splice enclosure while the installer makes the splice
- While not a rule of thumb, some outdoor splice enclosures require four to six feet of buffer tube per cable per end

6- Termination and Splicing 6-174

174

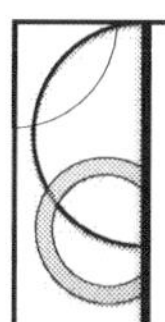

Outdoor Enclosures

- Single shell or multiple shells
  - Single shell most common type
- The shell can be a single piece or multiple pieces
  - Single piece shell has reduced concern for moisture sealing (opinion)
- Locations
  - Hang from cable
  - Pedestals

6- Termination and Splicing 6-175

175

Single Piece Outdoor Enclosure

(Courtesy Preformed Line Products)

6- Termination and Splicing 6-176

176

Multiple Piece Outdoor Enclosure

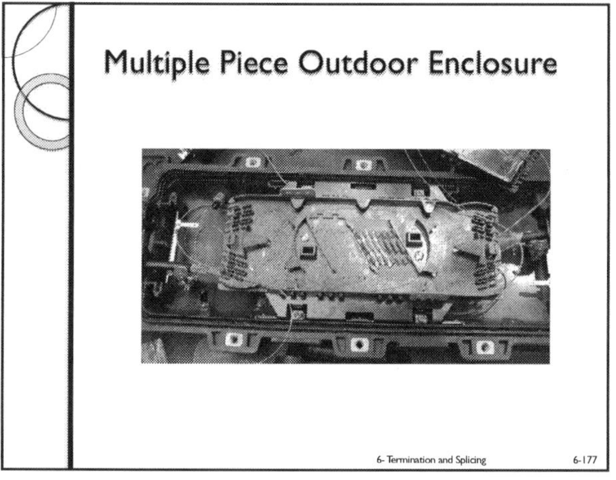

6- Termination and Splicing 6-177

177

Half Length Tray

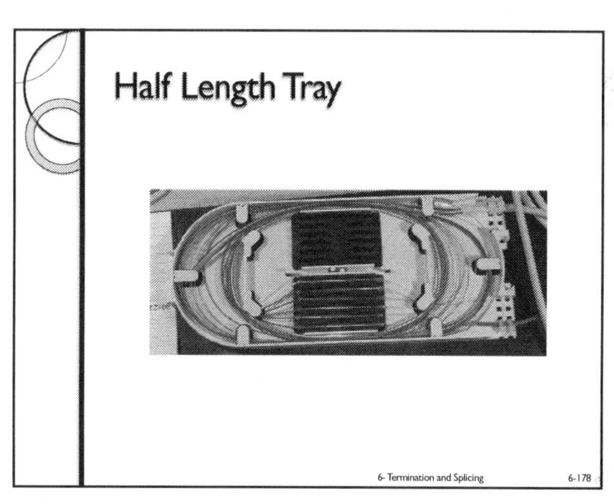

6- Termination and Splicing 6-178

178

Manhole With Enclosure

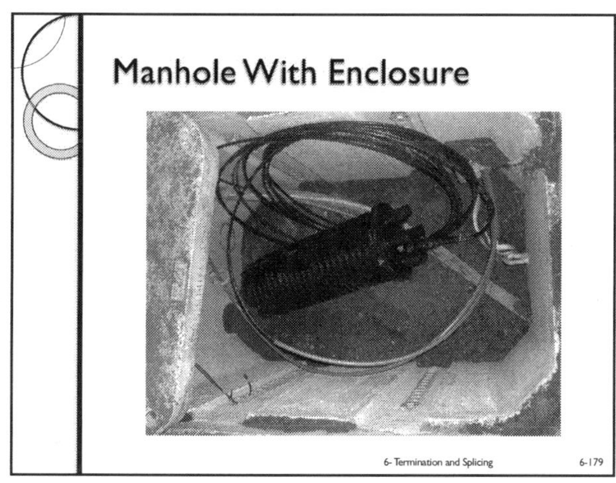

6- Termination and Splicing 6-179

179

7-Cable Port Enclosure

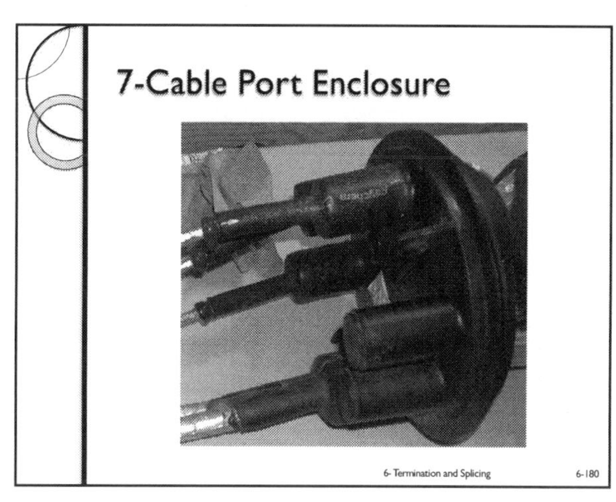

6- Termination and Splicing 6-180

180

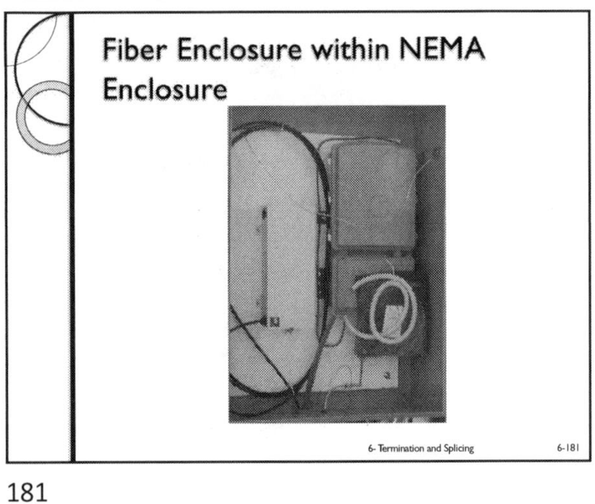

181

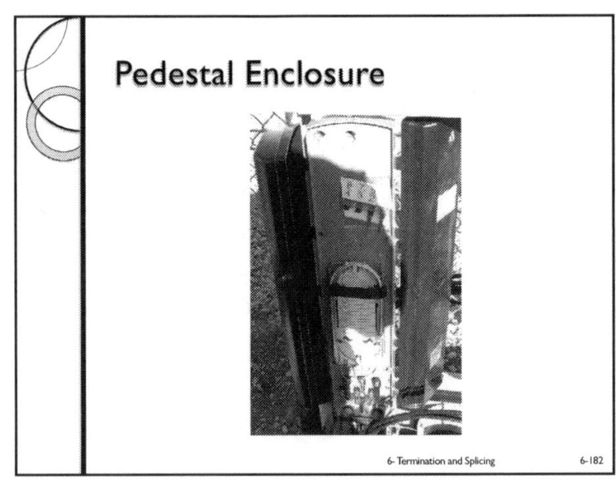

182

183

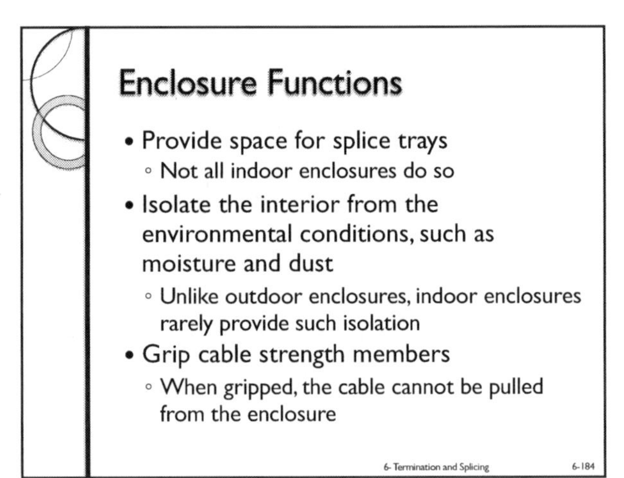

184

Indoor Enclosure Cable Gripping Mechanism

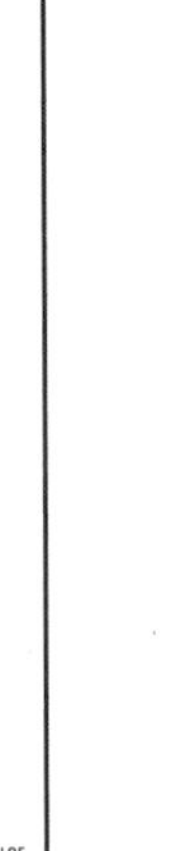

6- Termination and Splicing 6-185

185

Outdoor Enclosure Cable Gripping Mechanism

6- Termination and Splicing 6-186

186

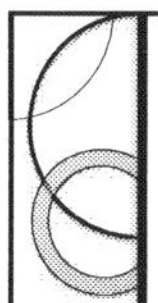

Secondary Components

- Moisture seals
- Gaskets
- Pressure valves
- Grounding strips
- Locking mechanisms
- Plugs

6- Termination and Splicing 6-187

187

Moisture Isolation

- Most outdoor, aerial enclosures provide seals for moisture isolation
- Such enclosures have a set of gaskets and grommets to prevent moisture ingress or internal pressure release
- Some outdoor enclosures provide mechanism for internal air pressure to verify proper gasket installation
  - With a soap bubble test

6- Termination and Splicing 6-188

188

Valve For Internal Pressure

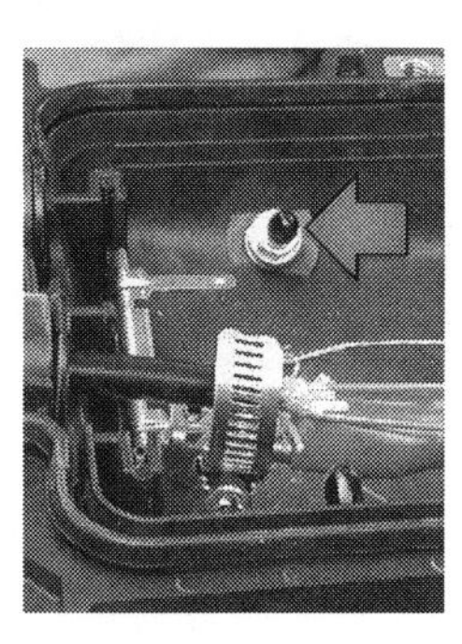

6- Termination and Splicing 6-189

189

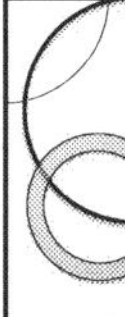

Integral Patch Panels

- Unlike most outdoor enclosures, indoor enclosures include an internal patch panel for direct connection of connectors to the fibers in the enclosure

6- Termination and Splicing 6-190

190

Essential Splice Information

- Definition
- Functions
- Locations
- Process
- Types
- Hardware
- Performance

6- Termination and Splicing 6-191

191

Performance

- Splice loss is less than 0.15 dB
- Both fusion and mechanical splices achieve this value, although fusion splices tend to have loss lower than that of mechanical splices
- Some organizations allow higher or lower values
- Examples
  - Some organizations have allowed splices as high as 0.5 dB.
  - Some organizations have required splices as low as 0.1 dB
- This author's experience
  - Most matched fibers produce typical loss of ~0.05 dB
  - Values above 0.1 dB are rare

6- Termination and Splicing 6-192

192

Splice Loss Options

- The Fiber Optic Association advanced splicing certification process requires all splices to be a maximum of 0.15 dB
- The Building Wiring Standard, TIA/EIA-568 D, allows splices to be up to 0.3 dB
- Recommendation: use 0.15 dB/splice or lower (opinion)

6- Termination and Splicing 6-193

193

Installer Needs to Know

- Type
- Mode type
- Maximum loss, in dB
- Typical loss, in dB
- Enclosure installation procedure

6- Termination and Splicing 6-194

194

Hands-On Mid-Span Splicing

- Do not proceed until you hear the magic word: Proceed
- Singlemode
  - Install splice cover
  - Strip, clean, cleave
  - Strip, clean, cleave second fiber
  - Install in splicer
  - Inspect cleaves
  - Splice
  - Shrink cover

6- Termination and Splicing 6-195

195

End Of Termination

Questions?
Comments?
Observations?

196

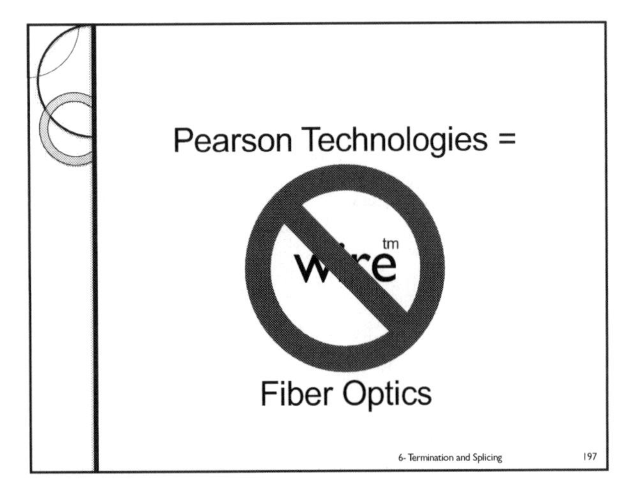

197

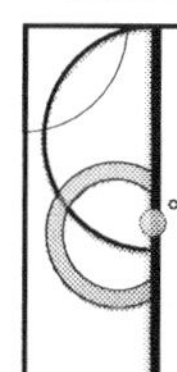

Professional Fiber Optic Installation

The Essentials For Success

Developed And Delivered By
Eric R. Pearson, CFOS
FOA Master Instructor
BICSI Master Instructor
Pearson Technologies Inc.

1

Before Building a Fiber Network, We..

DESIGN

7- Design 2

2

Design

- Definitions
- Design Steps
- Define Testing
- Documentation
- Budgets

7- Design 3

3

Definitions 1

- Cable plant is the light path from transmitter to receiver
  - Link is another term for 'cable plant'
- Accurate operation requires proper power level
  - Proper power level means sufficiently high and not excessively high
  - If power level is not proper, the specified bit error rate (BER) will be unacceptably high

7- Design 4

4

Definitions 2

- Loss budget is the calculated total loss of all components in the link
- Power budget is the maximum loss that can occur between a transmitter and receiver while functioning at the specified bit error rate (BER)
- Required minimum loss is the minimum loss required to avoid overloading a receiver and creating high BER

7- Design 5

5

Choose Cabling and Media

- Potential media types
  - Multimode, singlemode, coax, wireless, line of sight, and twisted pair
- While fiber is a dominant choice, it is not always the most cost-effective or best choice
  - Example: building management systems use proprietary copper cables
- Choosing media early in design process blinds the designer to finding the 'best' choice
- Bandwidth, distance, and cost combine to favor one technology more than another

7- Design 6

6

Twisted Pair?

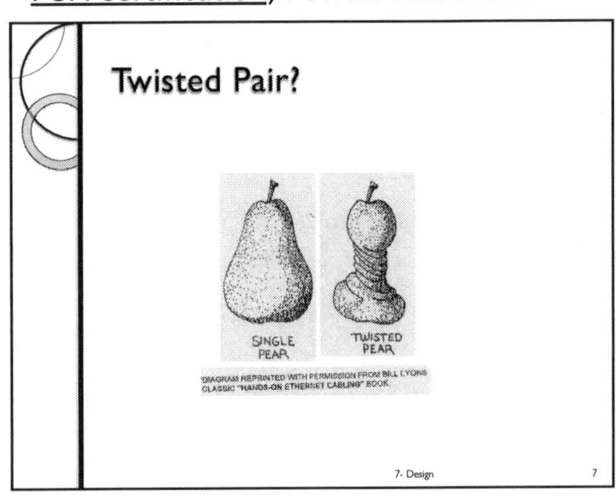

7- Design 7

7

Design

- Definitions
- Design Steps
- Define Testing
- Documentation
- Budgets

7- Design 8

8

Design Steps

- First considerations: user to defines application(s)
  - What data will network carry?
    - That is, transmission requirements or types of systems
  - Between what locations will network be built?
  - What will be the routes?
- Obtain the knowledge required
- Define connections and routes
- Choose cabling and media
- Define testing
- Create complete documentation

7- Design 9

9

Define Applications 1

- Data
  - Premises
    - Fiber cable commonly installed within building with multimode fiber for present use and with singlemode fiber for future upgrades
    - 50μ OM3 multimode fiber has the generally accepted minimum performance
    - Indoor cable standard provides common specifications
      - ANSI/ICEA S-83-596-2001
  - Building to building
    - Singlemode fiber most common
    - Outdoor cable standard provides common specifications
    - ANSI/ICEA S-87-640-2016
- CATV
  - Singlemode only

7- Design 10

10

Define Applications 2

- Access control
- Building management
  - Most such systems use proprietary copper cabling
- Security
- Telephone
- Metropolitan networks

7- Design 11

11

Metropolitan Networks Contain

- Surveillance cameras (CCTV)
- Emergency service communications
- Traffic monitoring and control
- Educational systems and services
- Telephone
- LAN
- Security
- Leased communication services on dark fibers

7- Design 12

12

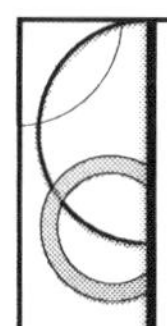

Define Transmission Requirements 1

- Define transmission types (types of systems)
  - Digital
  - Analog
    - CATV
- Define bandwidth required
  - At start
  - In future
    - Prime project manager has bandwidth needs
    - Other organizations may have transmission needs that can run on same transmission path and provide additional funding for project

7- Design 13

13

Define Transmission Requirements 2

- Finalize bandwidth requirement
- Specify protocol
  - Ethernet dominant for data transmission
    - Gigabit Ethernet
      - Multimode, 850nm, 1000Base-SX or
      - Singlemode, 1310nm, 1000Base-LX
    - 10 Gigabit Ethernet (10GBase-SX or 10GBase-LX)
  - Other

7- Design 14

14

Knowledge Requirements 1

- Applicable standards, including
  - TIA/EIA-568-E
    - Also known as The Building Wiring Standard for Copper and Fiber
  - FOTP-14, aka OFSTP-14, multimode insertion loss test procedure
  - FOTP-7. aka OFSTP-7, singlemode insertion loss test procedure
  - GR-771-CORE, requirements for outdoor enclosures
- Local codes
- Local regulations
- Electrical power systems
  - Electronics require stable and uninterruptable power

7- Design 15

15

Knowledge Requirements 2

- Telecom technologies
  - Cabled
  - Wireless
- Site surveys
- Fiber components and systems
- Installation processes
- Testing methods and interpretations

7- Design 16

16

Define Connections

- Working assumption: all end points need the same bandwidth
- Initially, connections define the end points of all transmissions
- After media chosen and installation methods reviewed, define intermediate connection locations

7- Design 17

17

Redundant Paths Needed?

- Mission critical networks require redundancy
- Two types available
  - Fully
    - Primary and back up fibers do not occupy the same space any where between two ends
  - Partial
    - Primary and back up fibers may occupy the same space between two ends
      - Same space can be: in same cable, on same side of street, in same conduit
      - Same space can be along part of path

7- Design 18

18

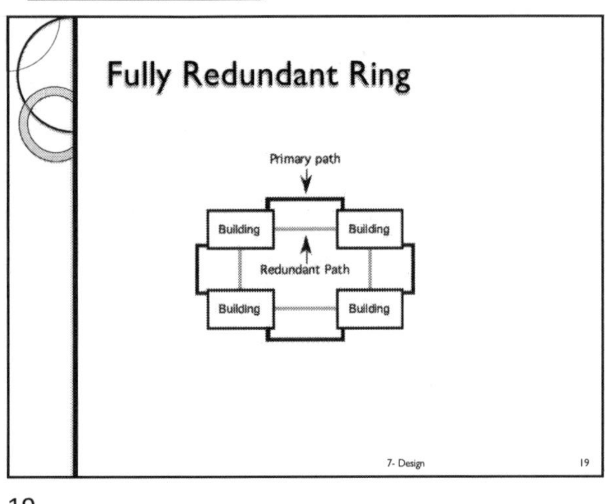

19

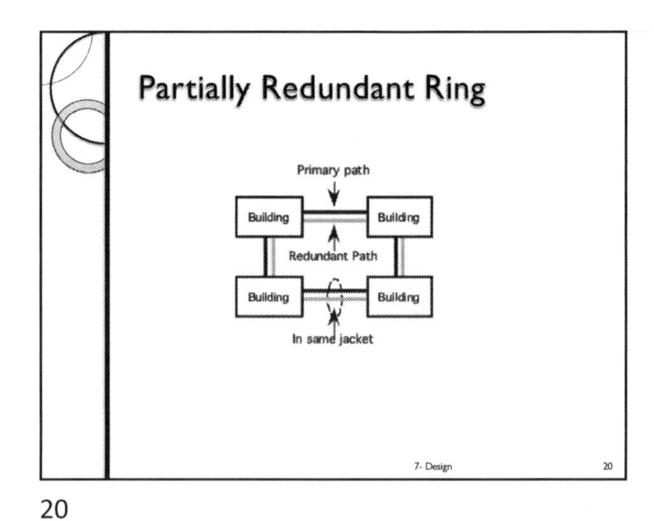

20

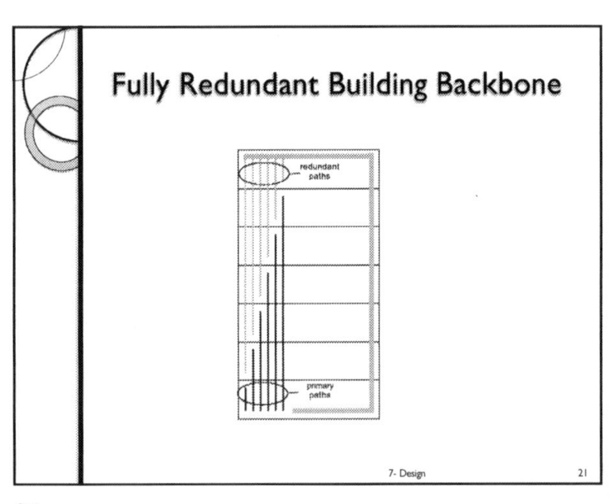

21

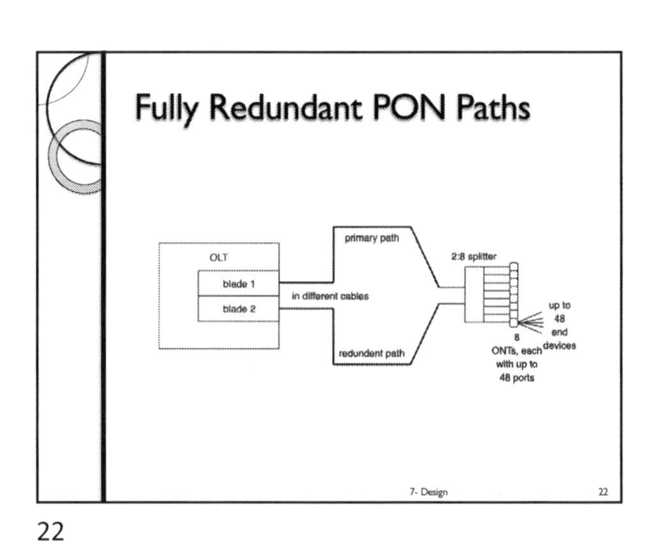

22

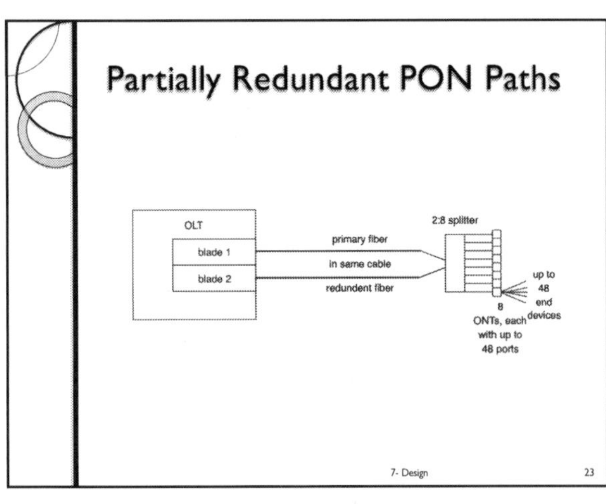

23

Map Identifies

- Cable paths
- Installation location options
  - Underground
    - Direct buried- require armor for crush resistance and/or rodent resistance
    - Installed in conduits- may require high installation load rating
  - Aerial
    - Lashed to messenger wire
    - Self support cable provides high long term use load
      - example all dielectric self support (ADSS)
- Connection locations
  - Drop locations
  - Splices
- Redundant paths

7- Design 24

24

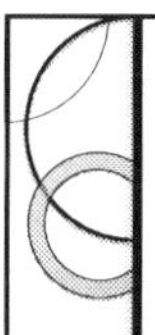

Cable Paths

- Cable path selection requires knowledge of
  - Local codes and regulations
- Path lengths are chosen
  - To comply with installation load limitations
  - To minimize connections
- End to end path lengths define transmission distances
  - Including service loops
- Since OSP cable lengths are often long, designer can order custom delivery lengths without increased cost
  - This approach minimizes splicing cost
  - If delivered lengths can be installed in single length

7- Design 25

25

Map Example

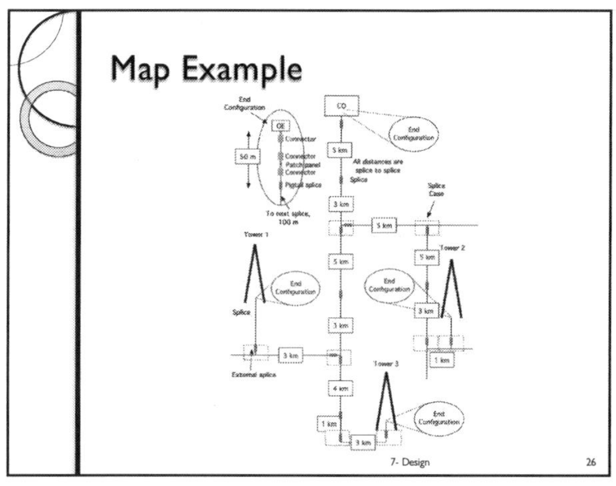

7- Design 26

26

Paths Determine Specifications for

- Environmental Requirements
- Installation Requirements

7- Design 27

27

Determine Cable Specifications

- Cable design
- Installation specifications
- Environmental specifications
- Legal
  - NEC requirements for indoor cables
- Size
  - Everything must fit
- Other
  - Color, markings

7- Design 28

28

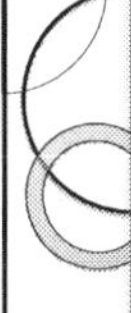

Commonly Used Cable Specifications

- ANSI/ICEA S-87-640-2016
  - Outdoor
- ANSI/ICEA S-83-596-2016
  - Indoor
- ANSI/ICEA S-104-696 (not in notes)
  - Indoor/outdoor
- Caution: review these these specifications to verify that all environmental and installation conditions in network are addressed

7- Design 29

29

Outdoor Cable Specifications

- ANSI/ICEA S-87-640-2016 provides these specifications
  - Singlemode
  - .4/.4/.4 dB/km at1310/1383/1550nm
  - Fiber: ITU-T G.652/G.657
  - Point discontinuity 0.1 dB
  - Operating temperature -40/+70°C
  - Storage temperature -40/+70°C
  - Installation-30/+60°C
  - Bend radius, minimum, 10x/20x
  - Installation load, maximum, 600 lbs-f 2670 N
  - Installation load, lower, 300 lbs-f 1330N
  - Plus moisture ingress, axial fiber strain <60% of fiber proof strain, usually 1%/100kpsi, compression

7- Design 30

30

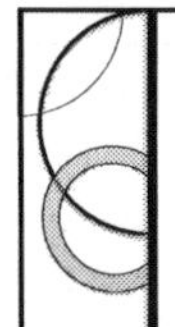

Underground Requirements

- Moisture resistance
- UV resistance
- Crush resistance

7- Design 31

31

Underground Options

- Direct buried cables require armor for crush and/or rodent resistance
- Installed in conduits
  - May require high installation load rating
  - Installation logistics may require
    - Reduction in cable segment lengths to avoid excessive installation load
    - Increased number of splices

7- Design 32

32

Aerial Installation

- Aerial Requirement
  - Moisture resistance
  - UV resistance
- Aerial Options
  - Lashed to messenger wire
    - Cables require standard installation load rating
  - Self support cable
    - Provides high long-term use load- up to 2000 lbs-f
    - Example: all dielectric self support (ADSS)

7- Design 33

33

G.652 Fiber Specifications

- MFD 8.6-9.5 nominal
  - MFD tolerance ±0.6μ
- Non circularity 1% maximum
- 0.69GPa/100kpsi proof stress
- Cladding 125μ
- Cladding tolerance ±1μ maximum
- Core offset 0.6μ maximum

7- Design 34

34

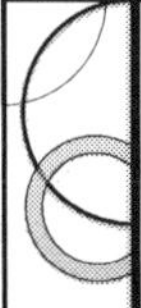

Environmental Limits

- Key principle: respect environmental limits
- Five common environmental limits
  - Moisture
  - Operating temperature range
  - Bend radii
  - Crush load
  - Use load or vertical rise distance
- Indoor cable requires NEC rating appropriate to location

7- Design 35

35

Moisture Problems

- All outdoor cables require moisture resistance
  - Channel moisture into electronics
  - Develop increased attenuation rate and fiber breakage due to frozen water
  - Experience reduced fiber strength and breakage due to attack from the chemicals in ground water
- If there is moisture in environment, cable must be moisture resistant;
  - I.e., dry water blocked with SAPs or
  - Gel-filled and grease-blocked

7- Design 36

36

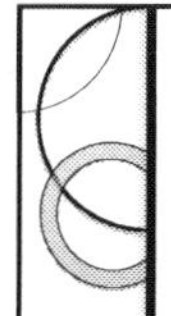

Operating Temperature Range

- Violation of range results in
  - Increased attenuation rate
  - Degradation of materials
- This is designer concern, but installer may need to know this range to explain high attenuation rate

7- Design 37

37

Generic Temperature-Attenuation Rate Performance

7- Design 38

38

Consider Installation Processes

- Ideal situation
  - Purchase cable as long as possible (long enough to span end points)
- Real situations
  - Cable reel may be too large to be handled by existing equipment
  - Conduit installation length may exceed installation load limit of desired cable with result that
    - Path will consist of multiple segments with splices
    - Splice location may be inconvenient for splicing resulting in increased service loop lengths

7- Design 39

39

Installation Limits (11-2)

- Respect installation limits
- Limits are on
  - Twisting- none allowed
  - Installation load- limit load
  - Installation bend radius- control radius
  - Installation temperature range- do not violate
  - Storage temperature range- do not violate

7- Design 40

40

If High Installation Load Required..

- Use loose tube design
- Remember the excess fiber length and mechanical dead zone?
- Example

7- Design 41

41

Bend Radii

- During and after installation, installer can bend cable no smaller than appropriate bend radii
- Violation results in
  - Increased attenuation rate
  - Fiber breakage
  - Reduction of fiber strength

7- Design 42

42

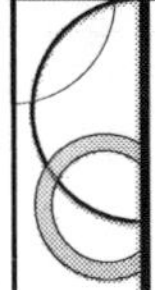

Bend Radius Rules

- Minimum bend radius, short term: 20x cable diameter
  - Defined under condition of maximum installation load rating
- Minimum bend radius, long term: 10x cable diameter
  - Defined at under condition of no load

7- Design 43

43

Crush Loads

- Two crush loads
  - Long term
  - Short term
- Violation results in
  - Increased attenuation rate
  - Fiber breakage
  - Reduction of fiber strength

7- Design 44

44

Use Load

- Long term, or use load, is limited
- If permanent load exists, it must not exceed rating of cable
- Installer must know this limit
- Violation results in
  - Increased attenuation rate
  - Fiber breakage
  - Reduction of fiber strength

7- Design 45

45

Second Form Of Use Load

- Maximum vertical rise distance
- Installer must limit vertical rise without mid-span support to rating of cable
- Data sheet will have this limit
- Install vertical cables from top down

7- Design 46

46

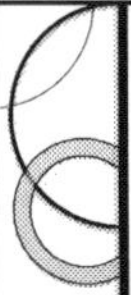

Indoor Cable Specifications

- ANSI/ICEA S-83-596-2016 (partial) provides these specifications
  - Singlemode
    - 1.0/1.0/1.0 dB/km at1310/1383/1550nm
    - Fiber: ITU-T G.652/G.657
  - Multimode
    - 305/1.5 dB/km at 850/1300nm
    - Fiber: ISO/IEC designation OM1, OM2, OM3, OM4
    - Minimum OM3 modal bandwidth: 1500 MHZ-km
    - Point discontinuity: 0.2 dB (multimode), 0.1 dB (singlemode)
    - Operating temperature, riser, general purpose: -20/+70°C
    - Operating temperature, plenum: 0/+70°C
    - Storage temperature: +40/+70°C
    - Installation temperature: riser, general purpose -10/+60°C
    - Bend radius, minimum, 10x/20x
    - Installation load, backbone, maximum, ≤12 fibers: 150 lbs-f 660 N
    - Installation load, backbone, maximum, ≥12 fibers: 300 lbs-f 1320 N

7- Design 47

47

Indoor Cable

- Requires compliance with local electrical code
- Local code usually references the NEC
- The NEC requires cable to be
  - Appropriate for location
  - Listed by Underwriters Laboratory (UL)

7- Design 48

48

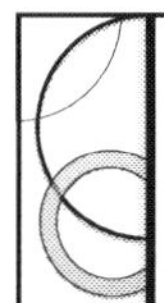

Ribbon Cable Favored for…

- High fiber count
- Large number of splices
- Space for cable is limited (conduits)
- Reason: ribbon splicing reduces splicing cost significantly

7- Design 49

49

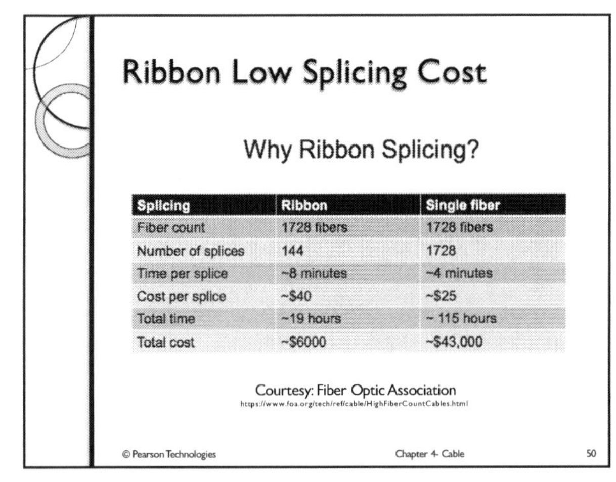

Ribbon Low Splicing Cost

Why Ribbon Splicing?

Splicing	Ribbon	Single fiber
Fiber count	1728 fibers	1728 fibers
Number of splices	144	1728
Time per splice	~8 minutes	~4 minutes
Cost per splice	~$40	~$25
Total time	~19 hours	~ 115 hours
Total cost	~$6000	~$43,000

Courtesy: Fiber Optic Association
https://www.foa.org/tech/ref/cable/HighFiberCountCables.html

© Pearson Technologies Chapter 4- Cable 50

50

Design

- Definitions
- Design Steps
- Define Testing
- Documentation
- Budgets

7- Design 51

51

Define Testing

- Insertion loss
- OTDR
- Microscopic Inspection
- Continuity and Polarity
- Documentation includes all test methods and results

7- Design 52

52

Insertion Loss Testing

- Performed after installation completed
- Performed with standard test method
  - OFSTP-7, singlemode
  - OFSTP-14, multimode
- Result compared to power budget
  - To verify electronics will 'work'
  - To verify that optical budget of electronics and loss of link are compatible
  - Loss or attenuation that is too high or too low results in high bit error rate (BER)
- Bidirectional testing can be performed
- Range testing advisable (author's opinion)

7- Design 53

53

OTDR Testing

- Bidirectional testing may be required
  - For true splice loss and stress loss
- Documentation includes all test results
- OTDR Testing may be performed three times
  - As-received cable (prior to installation)
  - After installation of each cable segment
  - After installation completed

7- Design 54

54

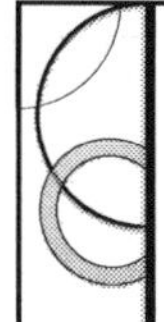

As-Received Cable Test

- Test verifies cable length and attenuation rate
- Test confirms lack of damage during shipment
- Visual inspection of sheath around reel may not indicate lack of damage

7- Design 55

55

Post Segment Installation Test

- When properly installed, cable will not exhibit breaks or increased loss
  - Breakage from
    - Excessive tension
    - Bend radius violation
  - Increased attenuation from violation of cable performance characteristic (e.g., bend radius)
- Verifies cable length and attenuation rate

7- Design 56

56

Design

- Definitions
- Design Steps
- Define Testing
- Documentation
- Budgets

7- Design 57

57

Cable Plant Documentation 1

- Documentation is critical and necessary part of design process but is often overlooked
- Documentation is most helpful for troubleshooting and restoration
- Documentation includes all information needed to facilitate installation and to implement upgrades and changes

7- Design 58

58

Cable Plant Documentation 2

- Documentation includes map that
  - Defines all optical paths
  - Identifies all unused ('dark') fibers
- All connection locations, types and losses
- Test results include
  - Insertion loss test results
  - OTDR values for lengths, attenuation rates, splice and connector losses
  - Location of all electronics

7- Design 59

59

Design

- Definitions
- Design Steps
- Define Testing
- Documentation
- Budgets

7- Design 60

60

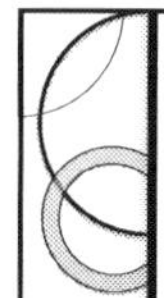

2 Budgets

- System power budget is the maximum loss that can occur between a transmitter and receiver while functioning below the specified bit error rate (BER)
- Determined by the protocol or standard to operate in network
- See examples in next slides
- Loss budget is the total loss of all components in the link

7- Design 61

61

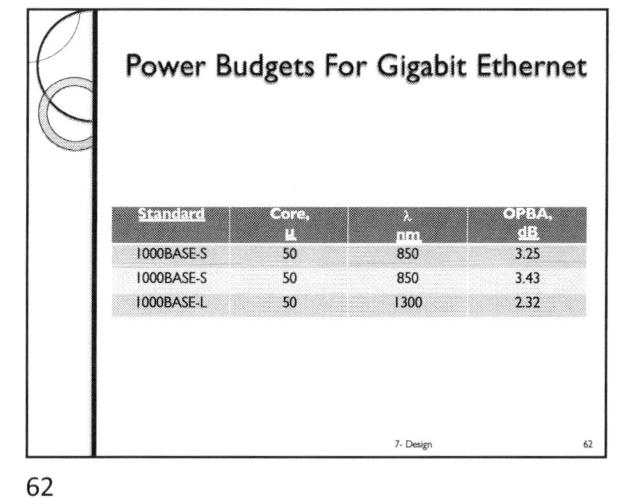

Power Budgets For Gigabit Ethernet

Standard	Core, μ	λ, nm	OPBA, dB
1000BASE-S	50	850	3.25
1000BASE-S	50	850	3.43
1000BASE-L	50	1300	2.32

7- Design 62

62

Power Budgets For 10GBASE-SX

	50	50	50	
Modal bandwidth	400	500	2000	MHz-km
Link budget	7.3	7.3	7.3	dB
Distance	66	82	300	m
Link loss	1.7	1.8	2.6	dB
Power penalty	5.1	5.0	4.7	dB

7- Design 63

63

Optional Power Budgets, 10GBASE-SX

No. Pairs	Distance, m	Loss, dB
2	300	2.77
3	280	3.28
4	250	3.92
5	220	4.45
6	180	5.02
7	120	5.62

7- Design 64

64

Power Budget, Singlemode 10GBASE-LX

Link power budget	9.4	dB
Transmission distance	10	km
Link loss	6.2	dB
Power penalty	3.2	dB

7- Design 65

65

Loss Budget 1

- Loss budget calculation is an essential part of design process, as it ensures plant component losses are appropriate to the electronics
  - i.e., link loss is less than loss allowed by electronics
  - Loss budget ≤ power budget
- Calculation is of total of loss of all optical components in cable path
  - Components include fiber, connectors, splices, and passive devices
  - Passive devises include couplers, splitters, wavelength division multiplexors and demultiplexors

7- Design 66

66

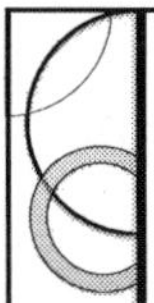

Loss Budget 2

- In the loss budget, the designer allows for measurement uncertainty
  - The generally accepted value is ± 0.2-0.5 dB
- Plant loss tests are compared to power budget to determine whether link will function properly
  - Stated otherwise, the losses are appropriate for the electronics

7- Design 67

67

Loss Budget 3

- Plant loss is determined with loss value for each component
  - Connectors on ends are included because that loss will be included in insertion loss test, as will be seen
  - Shorthand: end connectors are counted as connections (also known as 'pairs'). See next 2 slides.

7- Design 68

68

End Connectors Create Pairs

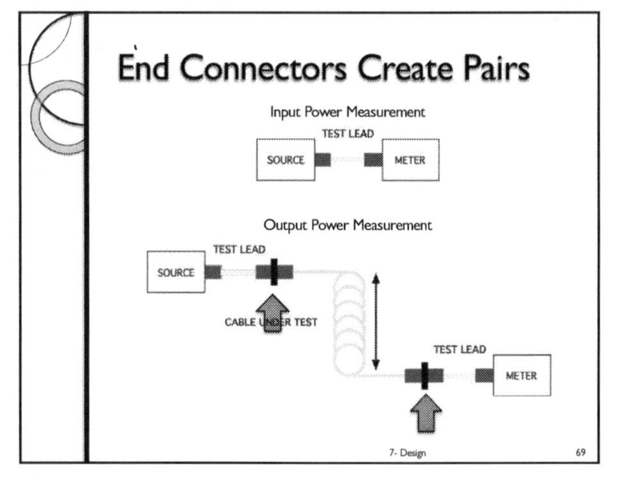

7- Design 69

69

Link Loss Includes End Pairs

TRANSMITTER

RECEIVER

LINK LOSS

7- Design 70

70

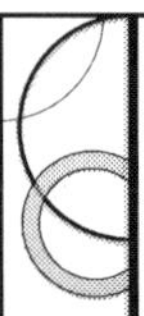

Which Component Loss Values to Use?

- Can use
  - Standard values
  - Typical loss
  - Designer-chosen values
  - Component manufacturer supplied values
  - Customer-chosen values
  - Values estimated from other installations
- Commonly, loss values chosen are conservative; that is slightly higher than expected or typical
- Can use any loss values as long as choice is included in documentation

7- Design 71

71

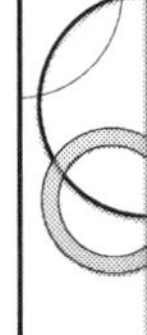

Choose Maximum Loss

- Can use maximum loss
  - As in industry standards
    - Example: TIA/EIA-568-E
- Industry maximums are generally recognized as being highly conservative
  - (the author's experience confirms this conservatism)

7- Design 72

72

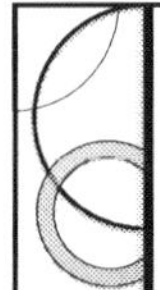

Choose Typical Loss

- Typical loss will be closer to actual loss than maximum loss
  - Source can be component manufacturer
- Disadvantage of using typical values
  - Even when properly installed, some actual values will be higher than typical
  - This case may defeat the purpose of loss budget

7- Design 73

73

Choose Designer Loss

- Designer-chosen values
  - Must be based on data, not guesses
  - Can be estimated values of typical losses
  - Example: telcos use such values
- Commonly, loss values chosen are conservative; that is slightly higher than expected or typical

7- Design 74

74

Map Determines

- Transmission distances
- But with multiple distances, need to select a strategy to design network
  - Reason: 20 links cannot result in 20 sets of specifications
- Choice of strategy will determine number of sets of specifications

7- Design 75

75

Choose Acceptance Loss Strategy

- Cables, connectors, and splices will have the same requirements throughout the network
- Optoelectronics may not have same requirements, if links have a wide range of lengths
  - Short links can work with low-cost optoelectronics
  - Long links may require high-cost optoelectronics

7- Design 76

76

Three Strategies Possible

1. Use maximum distance in network
2. Use maximum distances from each group for groups of links having approximately the same distance
3. Use the distance for each link

7- Design 4-77

77

Strategy 1: Maximum Distance

- Use maximum link length in network to determine optoelectronic specifications
- Advantages
  - One set of specifications
  - One set of products
  - One set of optoelectronic spares
  - Any optoelectronic works anywhere (almost)
- Disadvantage: potential for increased network cost

7- Design 4-78

78

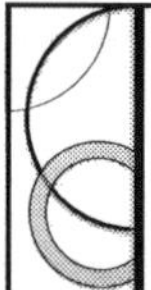

Example

- 30 links, one at 15,500', all others < 5000'
  - 15,500' link requires high-cost optoelectronics
- Use of same optoelectronics in all links increases cost
- All other links could use low-cost optoelectronics

7- Design 79

79

Strategy 1 Used

- Intra-building networks
- Telephone networks
- Inter-building networks if buildings closely spaced
- Networks in which any product must work everywhere
- Disadvantage: long link optoelectronics may overload receiver on short links

7- Design 80

80

Strategy 2: Grouped Distances

- Group links into groups of approximately the same length
- For each group, the designer determines the dispersion requirements
- Advantage
  - Pay for high price optoelectronics for links that need it
  - Pay for low price optoelectronics for those links that do not require high cost, high performance optoelectronics

7- Design 4-81

81

Disadvantage

- Multiple sets of spares

7- Design 82

82

Strategy 2 Used

- Inter building networks
- Example
  - 30 links, one at 15,500', all others < 5000'
  - 15,500' link requires high-cost optoelectronics
  - All other links can use low-cost optoelectronics
- Networks with redundant links

7- Design 83

83

Strategy Selection

- Designer selects strategy
- Policy may set strategy: one set of spares
- Cost consideration may determine strategy

7- Design 4-84

84

Attenuation Rates, Maximum

Wavelength, nm	Core Diameter, µ	Attenuation Rate, dB/km
850	50	3.0
1300	50	1.5
1310	8.2	0.5, 1.0
1550	8.2	0.5, 1.0

Note: 0.5 is a loose tube value
1.0 is a tight tube value

7- Design 85

85

Attenuation Rates, Typical

Wavelength, nm	Core Diameter, µ	Attenuation Rate, dB/km
850	50	2.0-2.3
1300	50	≤0.6
1310	8.2	0.30-0.35
1550	8.2	0.15-0.20

7- Design 86

86

Loss Calculation

- Choose type of value
  - Attenuation rate depends on wavelength and core diameter
- Calculate loss as in following slide
- Designer compares this value to power budget
  - This verifies electronics will function on link
- After testing, installer compares measured loss to calculated value

7- Design 87

87

Loss Budget Calculation (14-2)

Component	dB/km		#km	=	dB
Cable Loss=		*		=	
	dB/connection		#pairs		
Connector Loss=		*		=	
	dB/splice		#splices		
Splice Loss=		*		=	
	Totals		dB		
Passive device losses=				=	
			Total dB	=	

7- Design 88

88

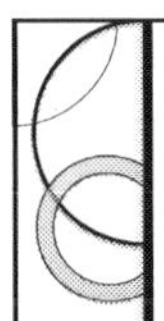

What Loss Will Work?

- 'Work' means that link operates at less than the specified bit error rate (BER)
- Work means that all component losses are appropriate for the electronics chosen
  - If loss is excessive/too high, the receiver does not receive sufficient power
  - If loss is insufficient/too low, the receiver receives an excessive power
  - In either case, improper loss results power level that prevents accurate conversion of optical signal to electrical signal

7- Design 89

89

Calculate Acceptance Loss Values

- Choose strategy
- If 'mid-point' strategy chosen, calculate insertion loss acceptance value as
  - (Maximum loss+ typical loss)/2
- If 'mid-point' strategy chosen, calculate OTDR acceptance values:
  - (Maximum attenuation rate+ typical attenuation rate)/2
    - Each segment must have uniform loss (straight backscatter)
  - (Maximum connector loss+ typical connector loss)/2
  - (Maximum splice loss+ typical splice loss)/2

7- Design 90

90

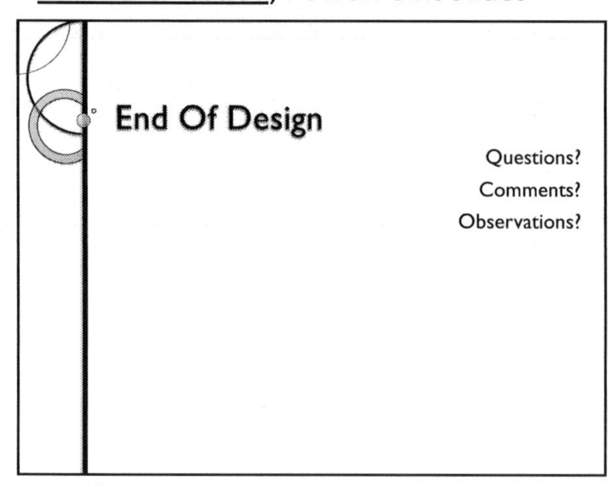
End Of Design
Questions?
Comments?
Observations?

91

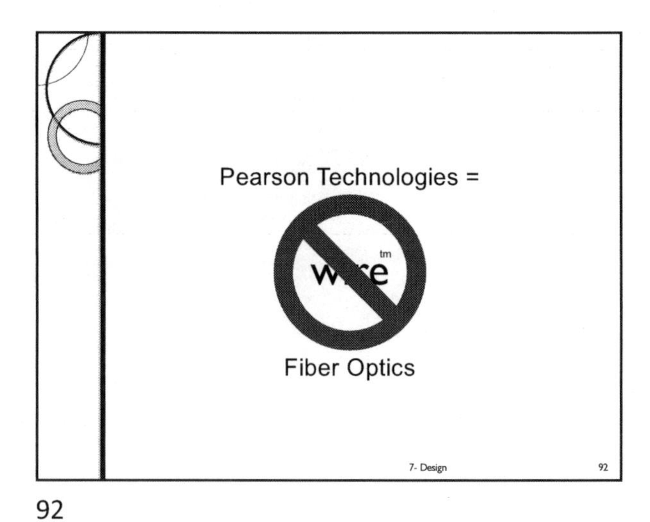
Pearson Technologies =
wire
tm
Fiber Optics
7- Design
92

92

Professional Fiber Optic Installation

The Essentials For Success

Developed And Delivered By
Eric R. Pearson, CFOS
FOA Master Instructor
BICSI Master Instructor
Pearson Technologies Inc.

1

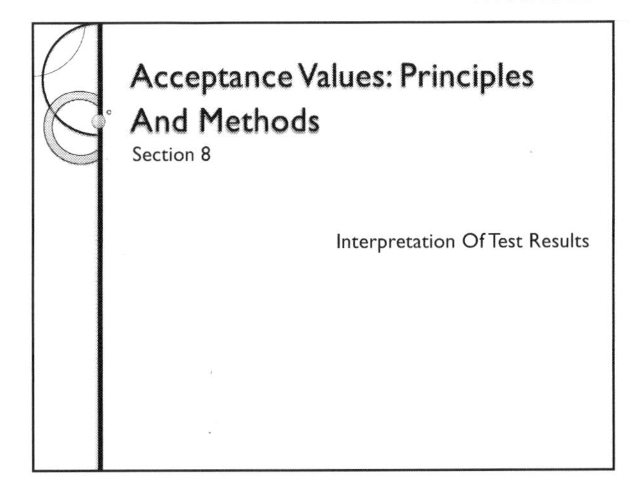

2

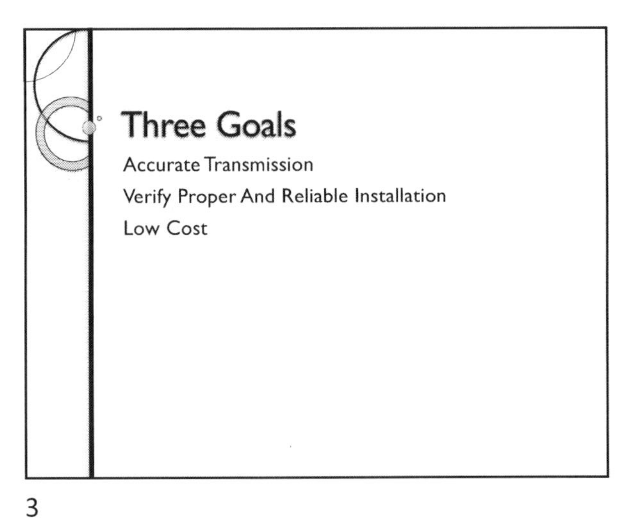

3

Certification Goals(19-1)

- Verify sufficiently low loss through the link
- Verify that each component in link has been properly and reliably installed

8- Acceptance Values: Principles And Methods 4

4

Applicability

- Calculation of acceptance values is performed when the differences between maximum and typical losses is large
- The designer makes the decision to use such values.
- These values are not the only values a designer can use

8- Acceptance Values: Principles And Methods 5

5

Two Acceptance Value Types

- Acceptance Values for
  - Insertion loss test values
  - OTDR test values

8- Acceptance Values: Principles And Methods 6

6

Necessary Step

- Choose acceptance value strategy from one of at least three strategies
  - Use maximum values
  - Use typical values
  - Use mid-point values

8- Acceptance Values: Principles And Methods 7

7

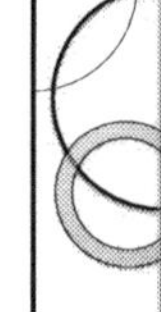

Four Steps

- Obtain required data
- Perform insertion loss calculations
- Calculate insertion loss acceptance values
- Calculate OTDR acceptance values

8- Acceptance Values: Principles And Methods 8

8

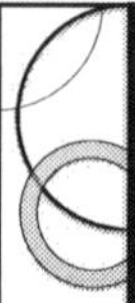

Required Data

- Accurate map
- Attenuation rate, maximum
- Attenuation rate, typical
- Connector loss, maximum
- Connector loss, typical
- Splice loss, maximum
- Splice loss, typical

8- Acceptance Values: Principles And Methods 9

9

Perform Insertion Loss Calculation

- Calculation depends on the Method used (Section 10)
- Use test method
  - Multimode Method B
  - Singlemode Method A.1
- Both treat two end connectors as two connections (pairs)

8- Acceptance Values: Principles And Methods 10

10

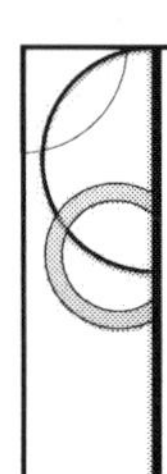

Maximum Loss Budget Calculation

Component	dB/km		#km	=	dB
Cable Loss=		*		=	
	dB/connection		#pairs		
Connector Loss=		*		=	
	dB/splice		#splices		
Splice Loss=		*		=	
	Totals		dB		
Passive device losses=				=	
			Total dB	=	

8- Acceptance Values: Principles And Methods 11

11

Typical Loss Budget Calculation (14-2)

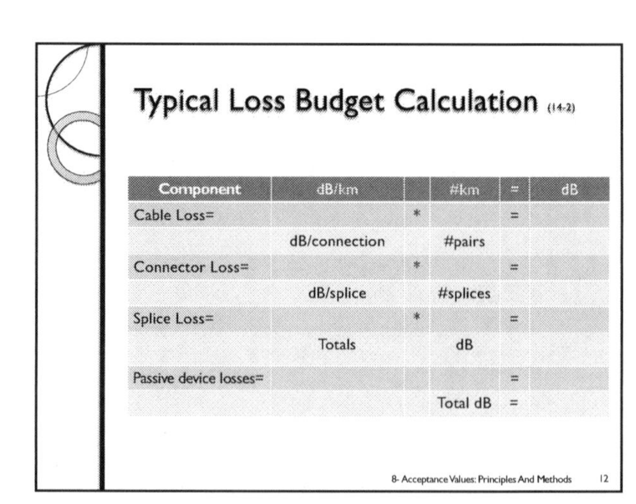

Component	dB/km		#km	=	dB
Cable Loss=		*		=	
	dB/connection		#pairs		
Connector Loss=		*		=	
	dB/splice		#splices		
Splice Loss=		*		=	
	Totals		dB		
Passive device losses=				=	
			Total dB	=	

8- Acceptance Values: Principles And Methods 12

12

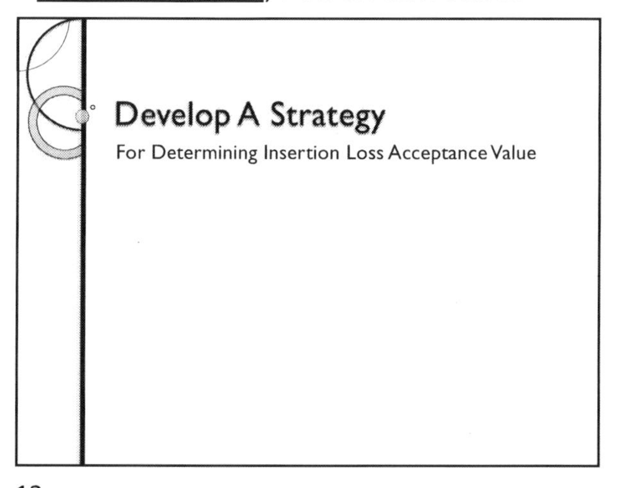

13

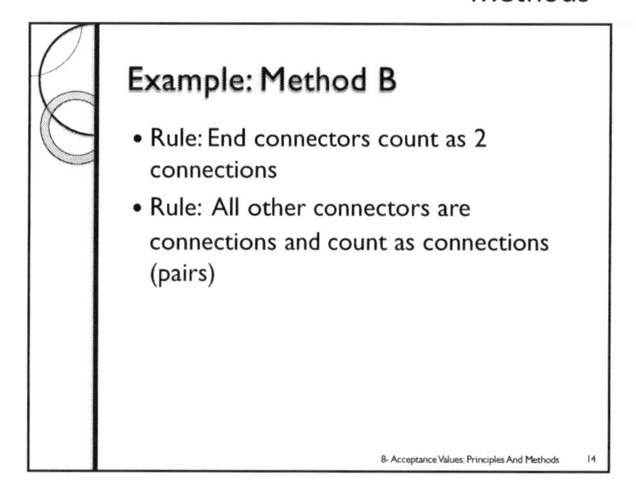

14

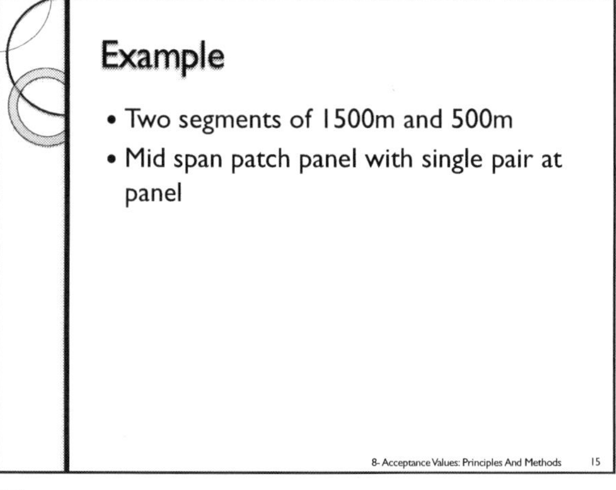

15

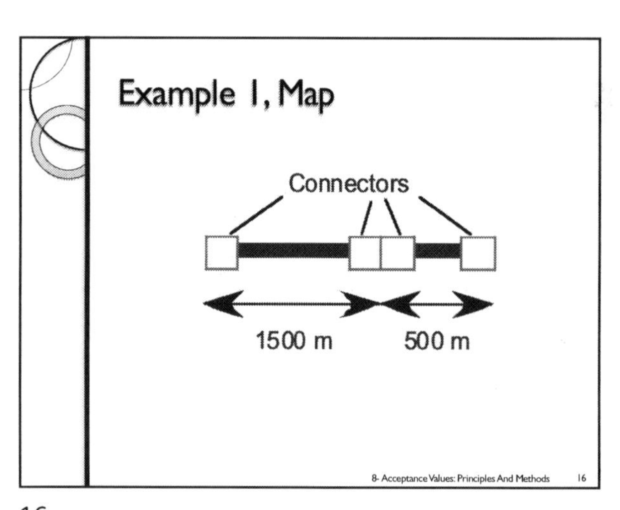

16

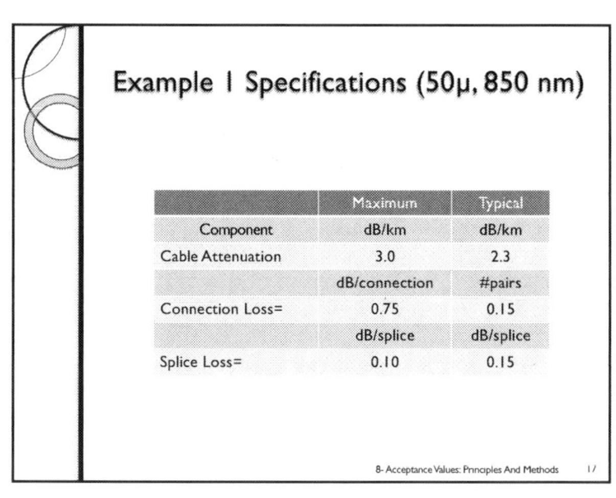

	Maximum	Typical
Component	dB/km	dB/km
Cable Attenuation	3.0	2.3
	dB/connection	#pairs
Connection Loss=	0.75	0.15
	dB/splice	dB/splice
Splice Loss=	0.10	0.15

17

Maximum Loss Budget Calculation

Component	dB/km		#km	=	dB
Cable Loss=		*		=	
	dB/connection		#pairs		
Connector Loss=		*		=	
	dB/splice		#splices		
Splice Loss=		*		=	
	Totals		dB		
Passive device losses=				=	
			Total dB	=	

8- Acceptance Values: Principles And Methods 18

18

Maximum Loss Budget Calculation

Component	dB/km		#km	=	dB
Cable Loss=	3.0	*	2.0	=	6.00
	dB/connection		#pairs		
Connector Loss=	.75	*	3	=	2.25
	dB/splice		#splices		
Splice Loss=	0	*	0	=	0
	Totals		dB		
Passive device losses=				=	0
			Total dB	=	8.25

8- Acceptance Values: Principles And Methods 19

19

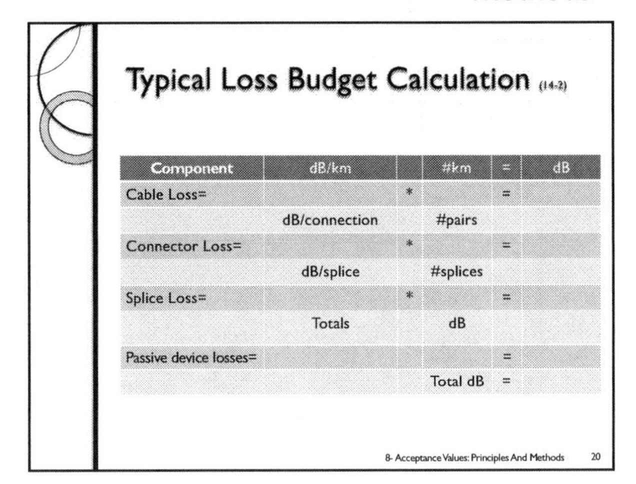

Typical Loss Budget Calculation (14-2)

Component	dB/km		#km	=	dB
Cable Loss=		*		=	
	dB/connection		#pairs		
Connector Loss=		*		=	
	dB/splice		#splices		
Splice Loss=		*		=	
	Totals		dB		
Passive device losses=				=	
			Total dB	=	

8- Acceptance Values: Principles And Methods 20

20

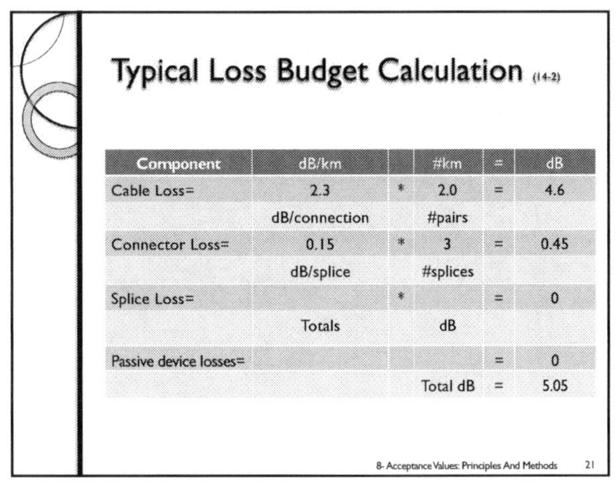

Typical Loss Budget Calculation (14-2)

Component	dB/km		#km	=	dB
Cable Loss=	2.3	*	2.0	=	4.6
	dB/connection		#pairs		
Connector Loss=	0.15	*	3	=	0.45
	dB/splice		#splices		
Splice Loss=		*		=	0
	Totals		dB		
Passive device losses=				=	0
			Total dB	=	5.05

8- Acceptance Values: Principles And Methods 21

21

What Happens If You Accept Maximum Loss?

- You allow up to 3.2 dB of excess loss.
- Excess loss occurs only through installation errors
- Installation errors reduce network reliability

8- Acceptance Values: Principles And Methods 22

22

Conclusion

- Do not want to use maximum loss as an acceptance value
  - Risk: reduced reliability

8- Acceptance Values: Principles And Methods 23

23

Alternative

- Use typical loss?

8- Acceptance Values: Principles And Methods 24

24

Typical Loss

- There is a normal variation of loss around the typical value
- Consider use of the typical value as a maximum acceptance value
- This use results in rejection of properly installed connectors and links that are slightly above typical values
- Benefit: none
- Disadvantage: increased cost

25

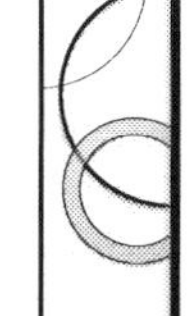

Conclusion

- Do not want to use typical loss as a maximum acceptance value
  - Risk: increased cost

26

How To Calculate Insertion Loss Acceptance Value

For High Reliability

27

Insertion Loss Acceptance Value

- Acceptance value is the highest insertion loss value that you will accept
- With a goal of high reliability
- Expect properly installed links to test closer to typical loss than maximum loss

28

Insertion Loss Acceptance Value

- Therefore, choose an acceptance value halfway between typical and maximum calculated values
- Call this the 'mid point acceptance value'

29

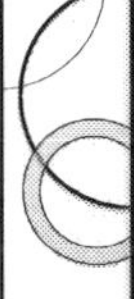

Five Part Strategy

1. Expect measured loss to be closer to calculated typical value than to calculated maximum value
2. Calculate mid point acceptance value
3. Accept Test Values ≤ mid point acceptance value
4. Investigate Test Values ≥ mid point acceptance value and ≤ calculated maximum value
   - Probably Reject
5. Reject Test Values > calculated maximum value

30

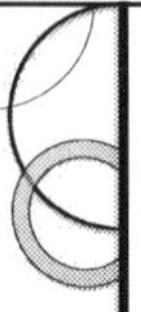

Investigate Means

- Test with OTDR
- Inspect connectors with microscope
- Inspect connectors and splices with VFL

8- Acceptance Values: Principles And Methods 31

31

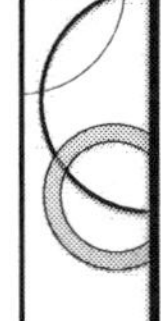

The Acceptance Value Is..

- Halfway Between Calculated Typical and Calculated Maximum Values
- = 1/2 * (5.05+8.25)= 6.65 dB
  - Use 6.7 dB

8- Acceptance Values: Principles And Methods 32

32

Mid-Point Acceptance Value

- Nothing magic
- This value allows minor mistakes

8- Acceptance Values: Principles And Methods 33

33

Example

- Multimode, 3M hot melt connectors are rated at typical loss of 0.3 dB /pair
- With over 10,000 installed in training and field work, this product never exceeds 0.4 dB /pair unless damage or dirt is visible on connector surface
- Mid point strategy allows connector loss up to 0.525 dB /pair
  - Key understanding: the mid-point acceptance value will accept minor mistakes

8- Acceptance Values: Principles And Methods 34

34

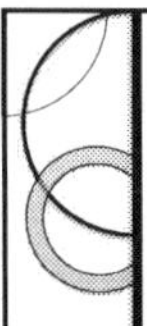

Calculate OTDR Acceptance Values

(19-3)

- Use same midpoint strategy to calculate acceptance values for OTDR connector losses, splice losses and attenuation rates
  - Accept test values less than or equal to the midpoint acceptance value
  - Investigate test values greater than the midpoint acceptance value and less than the maximum value
  - Reject test values greater than the maximum value

8- Acceptance Values: Principles And Methods 35

35

OTDR Acceptance Values

- Attenuation rate acceptance value= (maximum rate + typical rate)/2
- Connector pair acceptance value= (maximum loss + typical loss)/2
- Splice loss acceptance value= (maximum loss + typical loss)/2

8- Acceptance Values: Principles And Methods 36

36

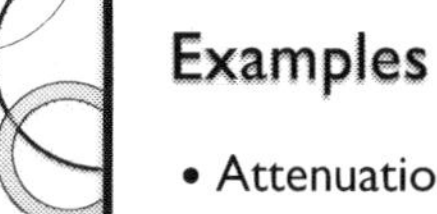

Example 1 Specifications (50μ, 850 nm)

	Maximum	Typical
Component	dB/km	dB/km
Cable Attenuation	3.0	2.3
	dB/connection	#pairs
Connection Loss=	0.75	0.15
	dB/splice	dB/splice
Splice Loss=	0.10	0.15

8- Acceptance Values: Principles And Methods 37

37

Examples

- Attenuation rate acceptance value= (3.0 + 2.3)/2= 2.65 dB/km
- Connector pair acceptance value= (0.75 + 0.15)/2= 0.45 dB/pair
- Splice loss acceptance value=
- (0.15 + 0.10)/2= 0.125 dB/splice

8- Acceptance Values: Principles And Methods 38

38

Additional Trace Requirement

- Every cable segment must exhibit a uniform loss (i.e., have a straight-line backscatter trace)
- A properly designed, properly manufactured, properly installed cable segment always exhibits a straight-line trace
- Any deviation from a straight-line trace indicates an installation error at the location of deviation

8- Acceptance Values: Principles And Methods 39

39

Perform Exercise 2

- Question 2, Section 19.7, Professional Fiber Optic Installation, v10
- Repeat Question 2 for two class links
- Determine insertion loss values
  - Maximum
  - Minimum
- Determine acceptance values
  - Insertion loss
  - OTDR connector loss
  - OTDR attenuation rate
  - OTDR splice loss

8- Acceptance Values: Principles And Methods 40

40

End Of Chapter

Questions?

Comments?

Observations?

41

Pearson Technologies =

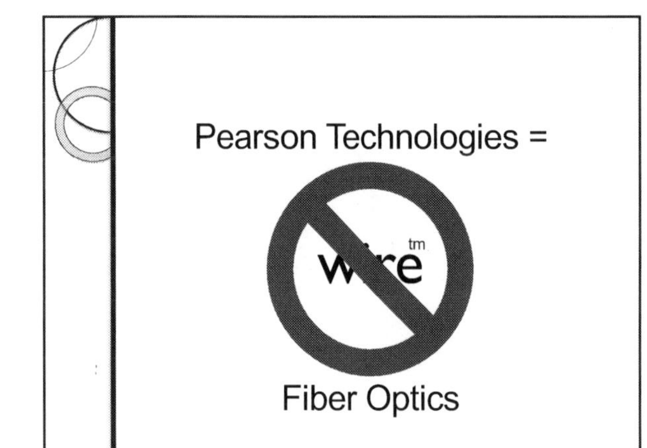

Fiber Optics

8- Acceptance Values: Principles And Methods 42

42

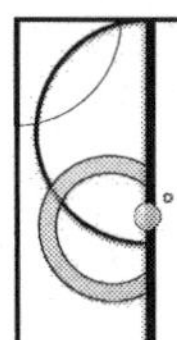

Professional Fiber Optic Installation

The Essentials For Success

Developed And Delivered By
Eric R. Pearson, CFOS
FOA Master Instructor
BICSI Master Instructor
Pearson Technologies Inc.

1

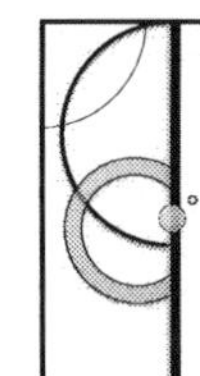

Three Program Objectives

One goal: be successful!

2

Objectives

- Success means
  - Low installation cost
  - Low power loss
  - High reliability
- Learn the basics of installation
- Learn the hands-on techniques for success
- Develop basic connector installation, inspection & testing skills
- Pass basic certification examination, Certified Fiber Optic Technician, or CFOT

9- Installation 3

3

Installation Concerns

- Goals
- Principles for Cable Installation
- Principles for End Preparation
- Planning and Management Considerations
- Safety Considerations

9- Installation 4

4

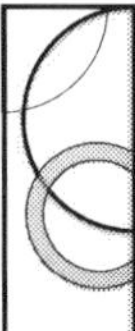

Four Installation Goals

- Avoid breakage
- Avoid reduced power at receiver
- Avoid reduced reliability
- Proceed safely

9- Installation 5

5

Installation Concerns

- Goals
- Principles for Cable Installation
- Principles for End Preparation
- Planning and Management Considerations
- Safety Considerations

9- Installation 6

6

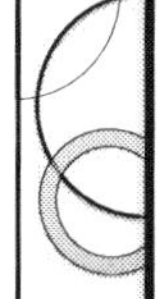

Cable Installation Principles

- Environmental Limits
- Installation Limits
- NEC Compliance
- End Preparation

9- Installation 7

7

Environmental Limits

- Key principle: respect environmental limits
- Five common environmental limits
  - Moisture
  - Operating temperature range
  - Bend radii
  - Crush load
  - Use load or vertical rise distance

9- Installation 8

8

Moisture Problems

- Channel moisture into electronics
- Develop increased attenuation rate and fiber breakage due to frozen water
- Experience reduced fiber strength and breakage due to attack from the chemicals in ground water

- If there is moisture in environment, cable must be moisture resistant; i.e., gel-filled and grease-blocked or dry water blocked with SAPs
- All outdoor cables need be moisture resistant

9- Installation 9

9

Operating Temperature Range

- Violation of range results in
  - Increased attenuation rate
  - Degradation of materials
  - 2 examples
- This is designer concern, but installer may need to know this range to explain high attenuation rate

9- Installation 10

10

Temperature Attenuation Rate Generic Behavior

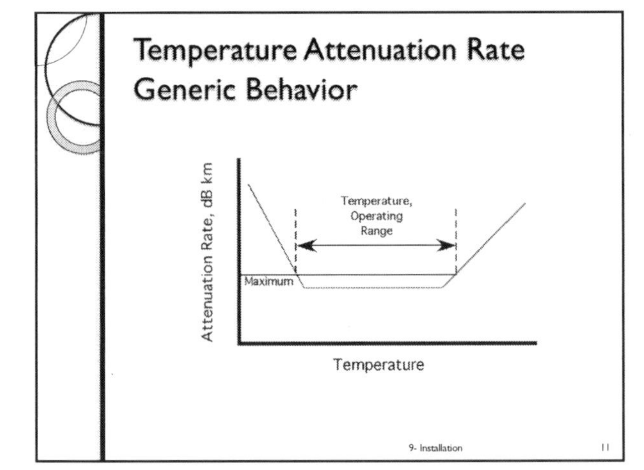

9- Installation 11

11

Bend Radii

- During and after installation, installer can bend cable no smaller than appropriate bend radii
- Violation results in
  - Increased attenuation rate
  - Fiber breakage
  - Reduction of fiber strength

9- Installation 12

12

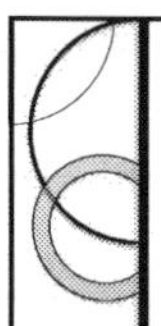

General Bend Radius Rules

- Short term: 20x cable diameter
- Long term: 10x cable diameter

9- Installation 13

13

Crush Loads

- Two crush loads
  - Long term
  - Short term
- Violation results in
  - Increased attenuation rate
  - Fiber breakage
  - Reduction of fiber strength
- See a pattern?

9- Installation 14

14

Use Load

- Long term, or use load, is limited
- If permanent load exists, it must not exceed rating of cable
- Installer must know this limit
- Violation results in
  - Increased attenuation rate
  - Fiber breakage
  - Reduction of fiber strength

9- Installation 15

15

Second Form Of Use Load

- Vertical rise distance
- Installer must limit vertical rise to value for cable
- Data sheet will have this limit

9- Installation 16

16

Cable Installation Principles

- Environmental Limits
- Installation Limits
- NEC Compliance
- End Preparation

9- Installation 17

17

Installation Limits (11-2)

- Respect installation limits
- Limits are on
  - Twisting- none allowed
  - Installation load- limit load
  - Installation bend radius- control radius
  - Installation temperature range- do not violate
  - Storage temperature range- do not violate

9- Installation 18

18

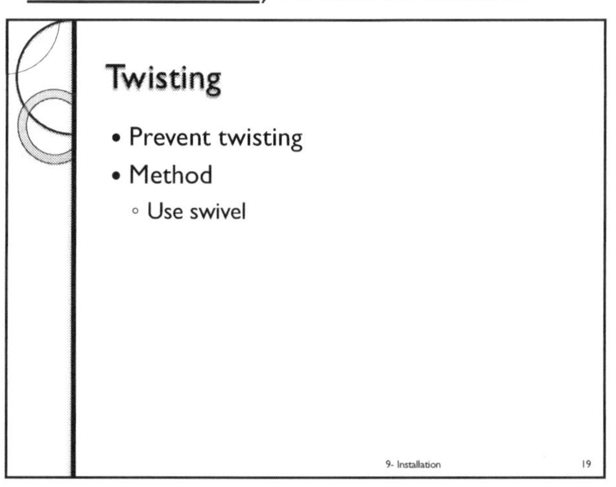

19

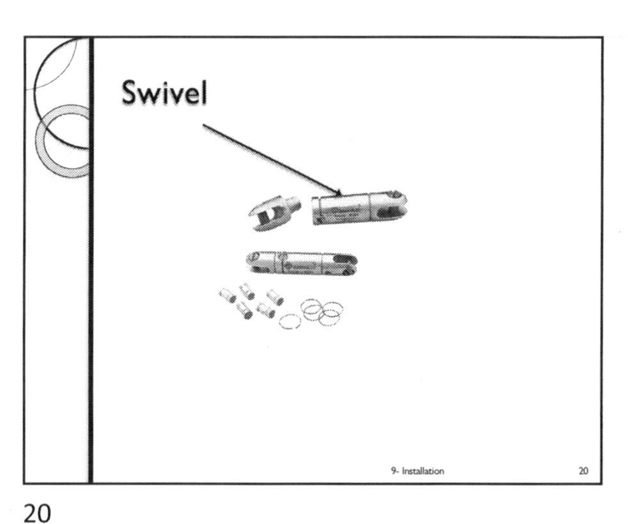

20

- Know load limit
- Limit load applied
- Three methods to limit load
  - Swivel with shear pin
    - Shear pin rated below cable installation load rating
  - Puller with slip clutch
    - Clutch set below cable installation load rating
  - Puller with load gage and overload shut off
    - Shut off level set below cable installation load rating

9- Installation 21

21

22

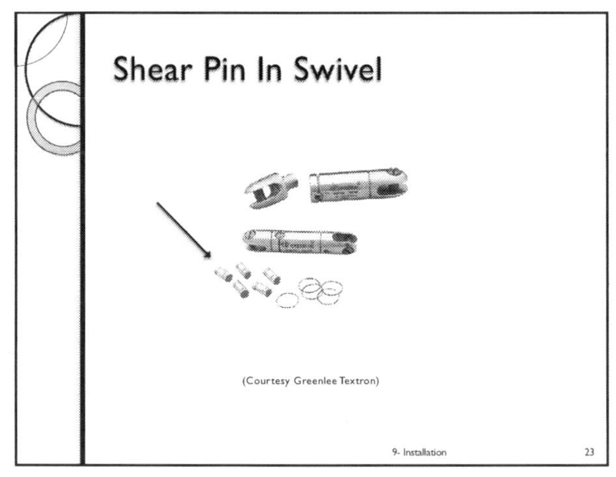

23

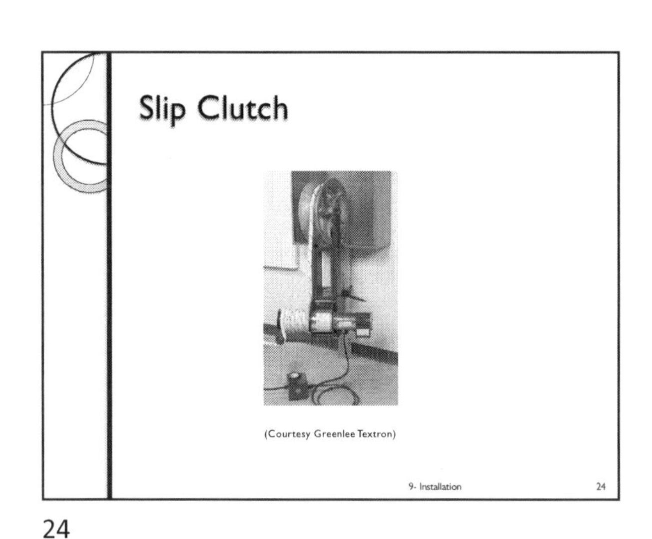

24

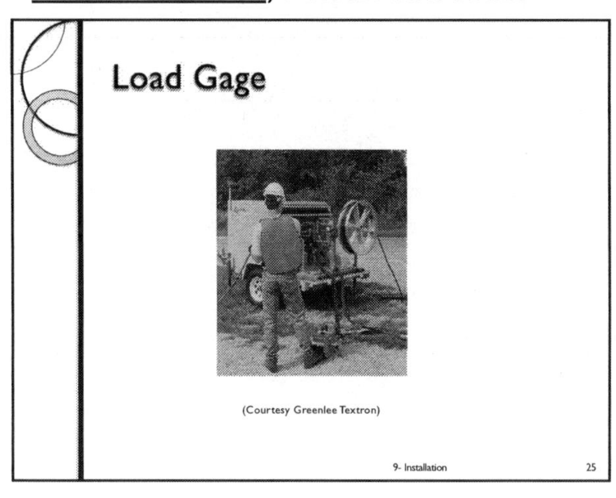

25

Reduce Load

- Use lubricant
  - Use fiber optic cable lubricant
  - Lubricant matched to jacket of cable
  - Fiber jackets are different from copper cable jackets
  - Fiber cable lubricant different from copper cable lubricant
- Use 'Figure 8' method

9- Installation 26

26

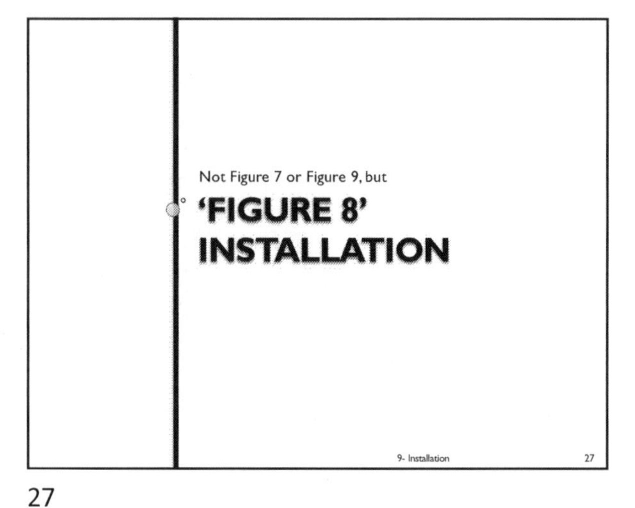

27

Benefits

- Reduce pulling load by reducing length pulled in
- Add half twist
- Half twist is removed in subsequent pull
  - Other wise, subsequent pull will put twist in cable
  - Remember the rule: no twisting

9- Installation 28

28

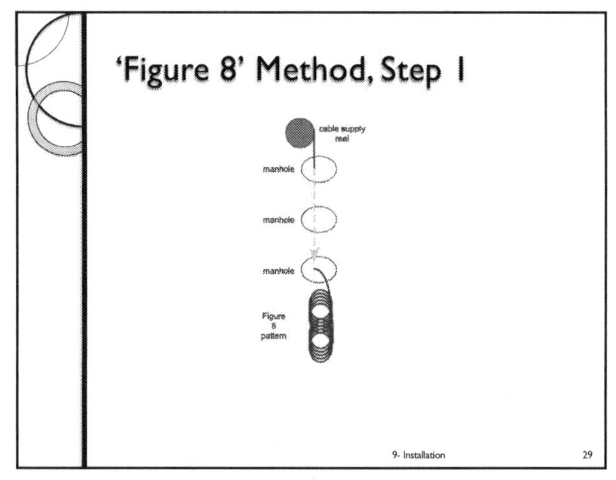

29

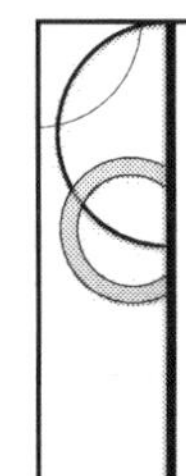

Figure 8 On Ground

9- Installation 30

30

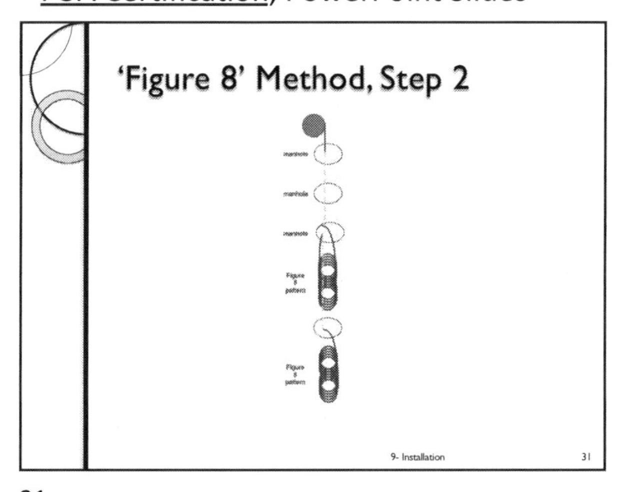

31

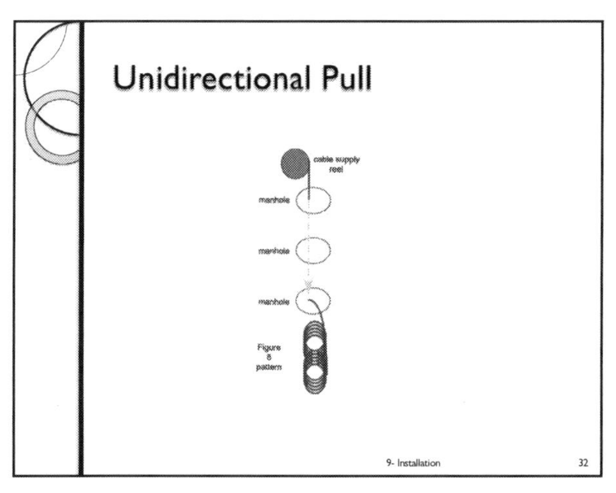

32

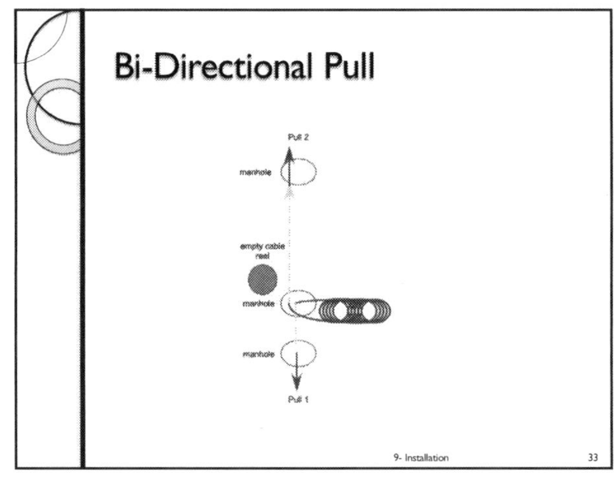

33

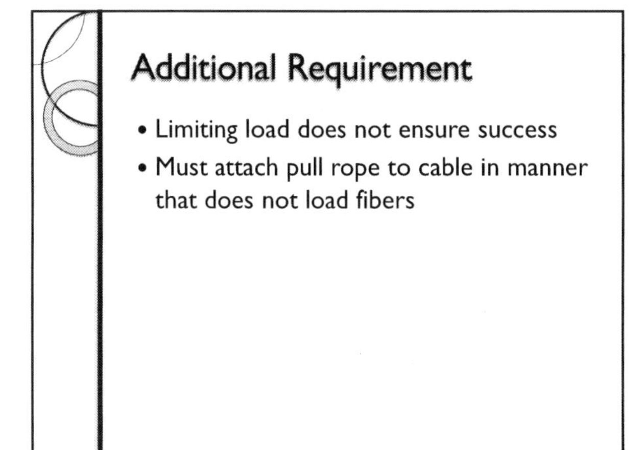

34

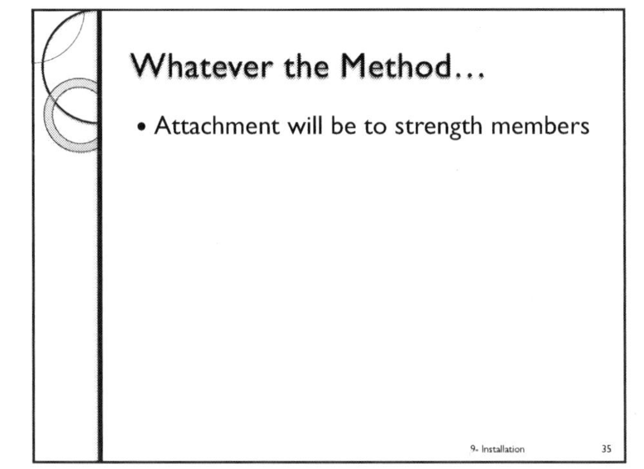

35

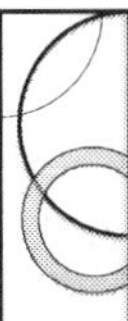

Attachment Methods

- Attachment of the pull rope around the outside of the cable jacket
- Attachment of the pull rope to a Kellems grip that grips the cable through the jacket
- Attachment of the pull rope to central strength members
- Attachment of the pull rope to strength members outside the loose buffer tubes
- Attachment of the pull rope to strength members between multiple jackets

9- Installation 36

36

Kellems Grip

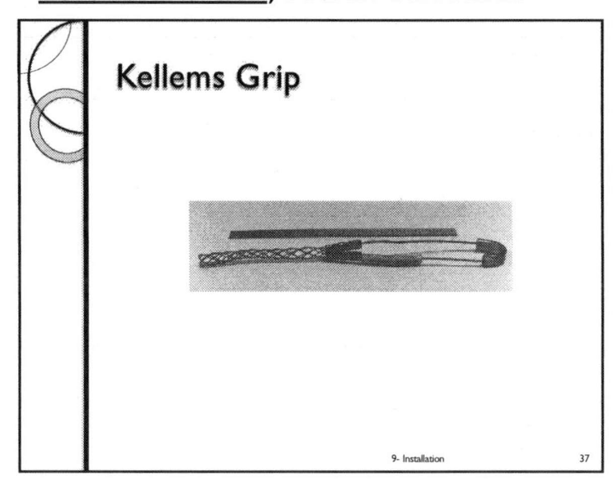

9- Installation 37

37

Which Method To Use?

- The method recommended by cable manufacturer
- Method may be in cable data sheet or application note

9- Installation 38

38

High Installation Load?

- Use loose tube design
- Remember the excess fiber length and mechanical dead zone?
- Example

9- Installation 39

39

Bend Radius

- Know minimum bend radius
- Limit bend radius
  - With sheaves or pulleys
- Minimum bend radius under load
  - 20x cable diameter
- Minimum bend radius without load
  - 10x cable diameter

9- Installation 40

40

Sheave

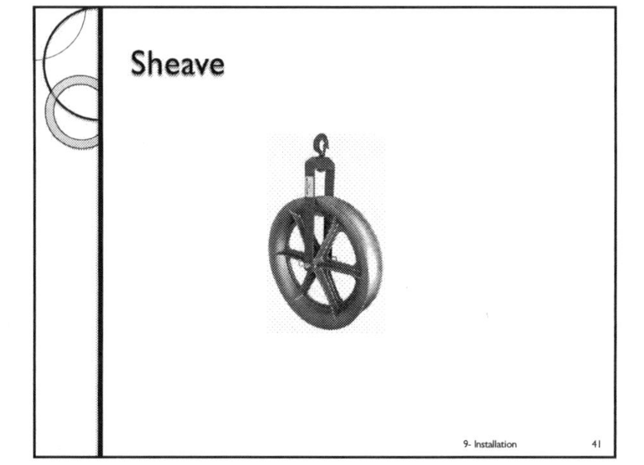

9- Installation 41

41

Monitor Supply Reel

- Monitoring avoids fiber breakage due to four conditions
  - Improper winding of cable on reel
  - Loosening of cable
  - Cable wrapping around the shaft supporting the reel
  - Back wrapping when the pull stops
- All can result in violation of bend radius

9- Installation 42

42

Improper Winding

43

Communication During Pulling

- Supply reel monitor must be prepared to stop reel as soon as puller stops pulling
- Otherwise, back wrapping and bend radius violation possible

44

Minimum Three People Per Pull

- One at reel
- One at puller
- One coordinating other two

45

Monitor All Pulleys To Prevent

- Cable jumping off pulley may cause bend radius violation
- Cable scraping sharp end of conduit and cutting into strength members under jacket
  - Then 600-pound force rating is lost and fiber breakage possible

46

Pull Cable

- Never push it
- Bend radius violation can occur
- In vertical runs, pull cable down, not up
  - This method reduces load applied to cable

47

Respect Temperature Ranges

- Two ranges
  - Installation
  - Storage
- Most important at temperature extremes
  - Arctic, desert
  - Examples
- Violation can cause
  - Jacket cracking, exposing fibers to environment
  - Degradation of cable materials
  - Increased attenuation rate

48

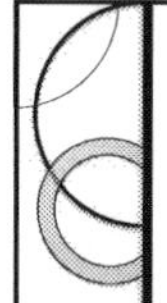

Install Outdoor Service Loops

- Reason
  - Service loops are inexpensive insurance
- Rules of thumb
  - 100' per 1000' of cable
  - 100' at street crossing
  - No more than 200' per 1000' of cable

9- Installation 49

49

Outdoor Aerial Cable

- Leave sag
- Sag allows for thermal expansion and contraction without stressing the fibers

9- Installation 50

50

Armored Cable

- Armored cables must be
  - Bonded at a mid span splices
  - Grounded at entrance to building and to equipment rooms
- Armored cable or any cable with a conductive structure must be grounded as it enters a building or wiring closet

9- Installation 51

51

Indoor Service Loops

- Reason
  - Service loops are inexpensive insurance
- Indoors
  - 6-10' at cable ends

9- Installation 52

52

Indoor Cable Placement

- Pull cable down in vertical riser
- Respect long term bend radius
- Bundle cables on same path
- Hand tighten cable ties
  - Use Velcro ties instead of cable ties
- Segregate fiber and copper cables
- Leave service loops everywhere
- Mark cable as fiber optic cable
- Comply with NEC requirements
  - Remove abandoned cables as required by NEC

9- Installation 53

53

NEC Compliance

- Do not violate the NEC
- Use cables with appropriate fire ratings
  - OFNP: install almost anywhere indoors
  - OFNR: install in riser and horizontal links
  - OFN: install only in links on same floor
- Remove abandoned cables
  - They increase the fire hazard

9- Installation 54

54

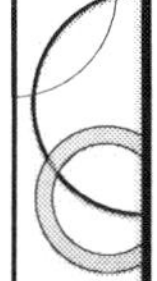

Connector and Dust

- During installation, dust is reported to cause 85-90% of problems
- Caps remain on connectors until connectors plugged into patch panel or electronics
- Barrels and electronics remain plugged until put into use

9- Installation 55

55

Installation Concerns

- Goals
- Principles for Cable Installation
- Principles for End Preparation
- Planning and Management Considerations
- Safety Considerations

9- Installation 56

56

Principles for End Preparation

- Remove jacket to length appropriate for enclosure
- Leave strength members exposed beyond end of jacket for attachment to enclosure
- Replace buffer tube with tubing or spiral wrap appropriate to enclosure
- Prepare cable ends so that buffer tubes and fibers remain inside of enclosure
  - Exception: break out cable
  - Buffer tubes are not designed to withstand exposure to working environment
- Seal ends of gel filled cable with fiber optic cable sealant

9- Installation 57

57

Cosmetics

- Some installers put heat shrink tubing over ends of cable to make end look neat or to prevent gel from seeping out of outdoor cables

9- Installation 58

58

Installation Concerns

- Goals
- Principles for Cable Installation
- Principles for End Preparation
- Planning and Management Considerations
  - Actions Prior To Installation
- Safety Considerations

9- Installation 59

59

Considerations 1-6

- Identify of the equipment and supplies required
- Determine the equipment locations
- Obtain data sheets for all products to be installed
- Identify the recommended installation techniques
- Determine the installation methods to be used
- Determine the people required

9- Installation 60

60

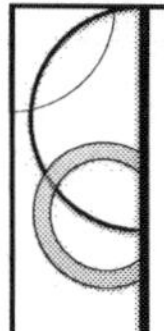

Equipment And Supplies

- Pulling equipment
- Pulling ropes
- Cable lubricant
- Pulleys
- Cable/Velcro ties
- Fusion splicer
- Insertion loss kit (OLTS)
- OTDR and launch cables
- Microscope
- Safety equipment
- Splice sleeves, fiber cleaner, tissues, gel/grease remover, batteries, VFL

9- Installation 61

61

Equipment Locations

- Pulling equipment may require anchor locations
- Connector installation may require power

9- Installation 62

62

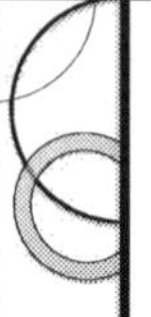

Data Sheets

- Installer requires data sheets for all products to be installed
- Data sheet information helps installer avoid unnecessary problems
- Examples
  - Cable data sheets include limits on installation load, use load, and bend radii
  - Mechanical splice data sheets identify cleave length

9- Installation 63

63

Installation Techniques

- Planner identifies techniques for installation of cables, connectors, and splices
- Application notes may include required information

9- Installation 64

64

Recommended Techniques

- Multiple installation techniques may be available
- Installer must choose technique best suited to installation conditions
- Data sheet or application note may identify best technique

9- Installation 65

65

Considerations 7-13

- Assign of activities to people with appropriate knowledge and experience
- Determine the testing requirements
- Create testing data forms
- Create as-built data logs
- Label cables, patch panel ports (TIA-606)
- Identify potential problem issues
- Identify potential safety issues

9- Installation 66

66

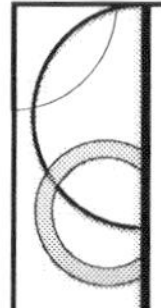

Personnel

- Supervisor determines number of installers required
- To avoid increased costs and reduced reliability, personnel must have training and experience appropriate to assigned activity
- Training can include certification (recommended) such as Fiber Optic Association certification (CFOT, CFOS)

9- Installation 67

67

Define Testing Requirements

- Prior to installation, supervisor defines testing needs
- After all link components installed, perform
  - Insertion loss testing
  - Microscopic inspection
- OTDR testing can be
  - On cable as received or before installation
  - After cable installation
  - After splicing
  - After all installation activities

9- Installation 68

68

As-Received Test On Cable

- Installer can perform an OTDR test of cable prior to installation
- Bare fiber adapter enables such testing
- Alternative: fusion or mechanical spliced pigtail
- Test verifies proper length and attenuation rate
- If no test prior to installation, how do you determine whether cable or installation was improper?

9- Installation 69

69

SC Bare Fiber Adapter

Courtesy Netceed

9- Installation 70

70

If No OTDR Test

- VFL test will indicate continuity

9- Installation 71

71

Post Installation OTDR Test

- Indicates lack of breakage
- Indicates no stress on fibers
- Alternative: a VFL continuity test
- Compare test cost to cost of finding break after connector installation

9- Installation 72

72

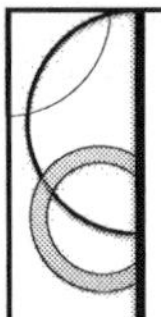

Post-Splicing Test

- Highly recommended
- Test splice after it is made and before closing enclosure
- Compare cost of test to cost of going to and opening enclosure to fix problem

9- Installation 73

73

Post Termination Test

- Final tests performed after all installation activities completed
- These tests enable certification and acceptance of link
- Documentation contains all results of tests and inspections

9- Installation 74

74

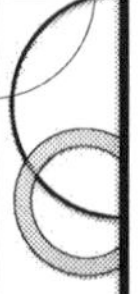

Data Forms

- Prior to installation start, create data forms
- Forms include
  - Test technique and source location
  - Wavelength
  - Spectral width
  - Equipment used
  - Direction of test
- Form enables duplication of test in future
- With duplication, changes will be due to changes in link, not to changes in test procedure

9- Installation 75

75

Documentation

- Final testing creates 'as-built' documentation
- Multiple copies of documentation recommended
- Documentation includes range test results
  - Range value enables proper interpretation of increases in loss

9- Installation 76

76

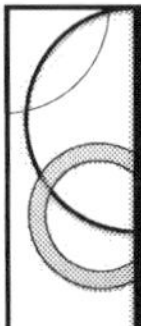

Potential Problem Identification

- Identify potential problems or problem areas
- Develop solutions to such potential problems
  - Including spare parts or equipment to resolve problem

9- Installation 77

77

Installation Concerns

- Goals
- Principles for Cable Installation
- Principles for End Preparation
- Planning and Management Considerations
- Safety Considerations

9- Installation 78

78

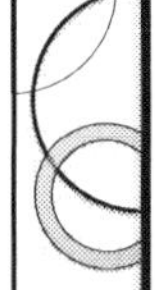

Identify Safely Issues

- During cable installation
  - Installation on active roadway may require lane shut down, flagmen, signs, etc.
- During splicing
- During connector installation
  - Keep unauthorized personnel out of work area
  - Dispose of bare fiber properly
- Bare fiber creation requires work rules regarding eyewear, food, and drink

9- Installation 79

79

Eye Safety 1

- Whenever the installer generates bare fiber, he wears safety glasses
- Imagine trying to find a 1/16" by 0.005" piece of glass in your eye:
  - How would you find it?
  - How would you remove it?
- Best policy: eliminate possibility of problem

9- Installation 80

80

Eye Safety 2

- Before looking into a connector with a microscope, check opposite end
  - There should not be active equipment attached
  - Light in fiber is invisible infrared wavelengths
  - Eye surgery by communication laser is not part of a normal fiber optic installation!
- Lock out or label opposite end
- Telephone links and FTTH links can have substantial power due to multiple wavelengths on same fiber

9- Installation 81

81

Eye Safety 3

- Identify local medical facility that has ability to locate and remove bare fiber from eye

9- Installation 82

82

Hand Safety

- Dispose of bare fiber as soon as you create it
- Use a fiber work mat that makes finding fiber splinters easy
- Prior to using rest room, wash hands

9- Installation 83

83

Clothing Safety

- Wear a 'fiber' smock that is washed separately from other clothing
- Wear smooth clothing
- After each session, use a fiber finding roller to ensure no fiber on clothing
  - Use from shoulders to knees
- Isolate work area: keep unauthorized personnel out of area in which you are creating bare fiber splinters

9- Installation 84

84

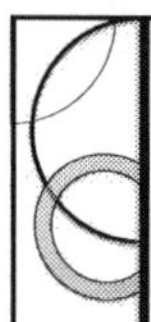

Mouth Safety

- No food, drink, smoke in area of work
- Fiber optic installation is not the same as a high fiber diet

9- Installation 85

85

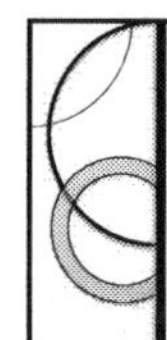

Chemical Safety

- Have Material Safety Data Sheets (MSDS) for all chemicals used
- MSDS has all information you need in an emergency

9- Installation 86

86

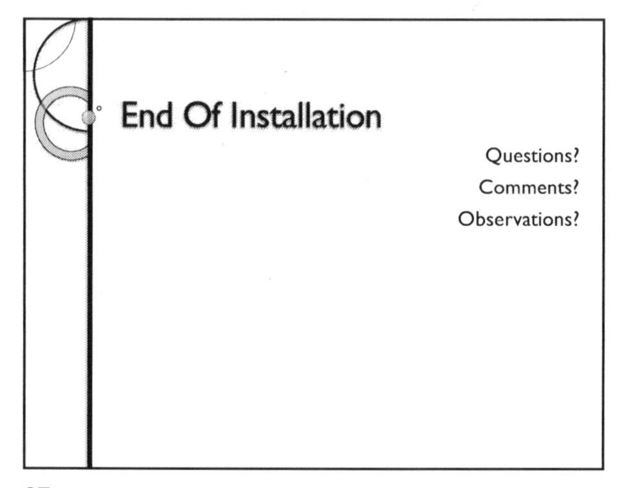

87

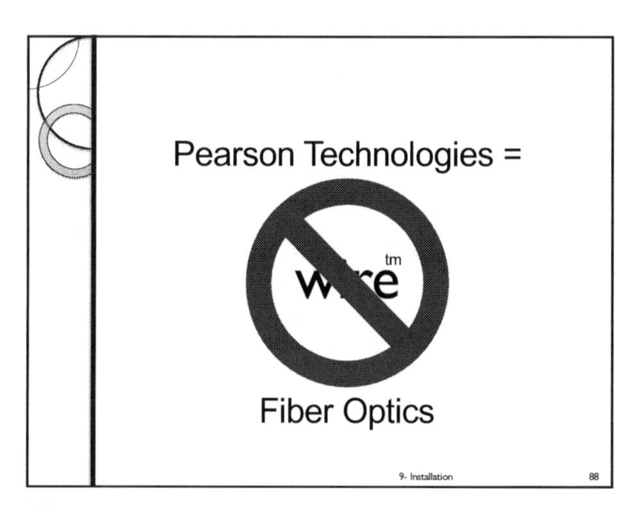

88

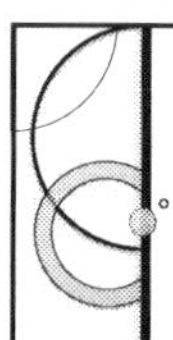

Professional Fiber Optic Installation

The Essentials For Success

Developed And Delivered By
Eric R. Pearson, CFOS
FOA Master Instructor
BICSI Master Instructor
Pearson Technologies Inc.

1

Testing and Inspection

- Goals
- Part 1 Insertion Loss
- Part 2 OTDR Testing
- Part 3 Microscopic Inspection
- Part 4 Continuity and Polarity Testing

© Pearson Technologies 10- Testing and Inspection 2

2

Goals: Verification of

- Accurate Transmission Through Insertion Loss Testing
- Reliable Operation Through OTDR Testing
- Low Loss Connectors Through Microscopic Inspection
- Continuity and Correct Polarity with VFL

© Pearson Technologies 10- Testing and Inspection 3

3

Testing and Inspection

- Testing Goals
- Part 1 Insertion Loss
- Part 2 OTDR Testing
- Part 3 Microscopic Inspection
- Part 4 Continuity Testing
  - Requirements

© Pearson Technologies 10- Testing and Inspection 4

4

To verify link will function properly, perform

INSERTION LOSS TESTING-PART 1

© Pearson Technologies 10- Testing and Inspection 5

5

Part 1: Insertion Loss Testing

- Goals
- Insertion Loss Testing
  - Singlemode
  - Multimode
  - Range Testing
  - Standards
  - Test at Two Wavelengths
  - Bi-Directional Testing
  - Equipment Requirements

© Pearson Technologies 10- Testing and Inspection 6

6

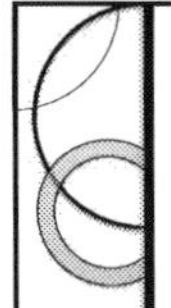

Key Concepts 1

- During transmission, two pulse changes occur
  - Dispersion
  - Power loss
- Dispersion testing not required
  - For data networks
  - Dispersion controlled by distance limitation and characteristics of fiber and transmitter
  - Dispersion cannot be increased by installation errors
- Power loss can be increased by installation errors
- Power loss testing always required

© Pearson Technologies 10- Testing and Inspection 7

7

Key Concepts 2

- Insertion loss testing is performed to verify loss is 'proper'
  - Not too high
  - Not not too low
- Insertion loss testing is performed after all link components are installed
- Insertion loss testing is always required, as it simulates the power loss that occurs between a transmitter and receiver
  - OTDR testing does not simulate this loss
- Insertion loss results are compared to loss estimated during project design phase

© Pearson Technologies 10- Testing and Inspection 8

8

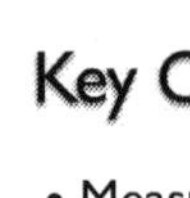

Key Concept 3: Test Principle

- Measure difference between power at input (transmitter) and power at output (receiver)
- Two measurements required
  - Input power
  - Output power
- Two measurements closely simulate loss in link

© Pearson Technologies 10- Testing and Inspection 9

9

3 Input Loss Measurement Methods

- One lead
  - Used to simulate link loss
  - Required by TIA/EIA-568
  - Provides smallest uncertainty
- Two leads
  - Method used when loss of end connectors not critical
- Three leads
  - Used for plug and jack connector types

© Pearson Technologies 10- Testing and Inspection 10

10

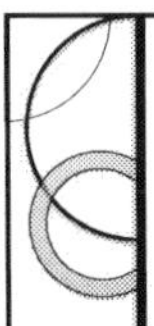

Part 1: Insertion Loss Testing

- Goals
- Insertion Loss Testing
  - Requires a source and power meter, aka 'optical loss test set, 'OLTS' (recently added)
  - Singlemode
  - Multimode
  - Range Testing
  - Standards
  - Test at Two Wavelengths
  - Bi-Directional Testing
  - Equipment Requirements

© Pearson Technologies 10- Testing and Inspection 11

11

Singlemode Testing

- Test with a laser diode
  - Wavelengths: 1310nm and 1550nm
- Measure input power with single test lead
  - Single test lead enables loss measurement to closely match link loss experienced by transmitter-receiver pair
- Single lead is Method A.1 of EIA/TIA-526-7
  - Aka OFSTP-7
- Link has at least 2 connections/connector pairs
- Measurement is of link loss including 2 connection/connector pairs

© Pearson Technologies 10- Testing and Inspection 12

12

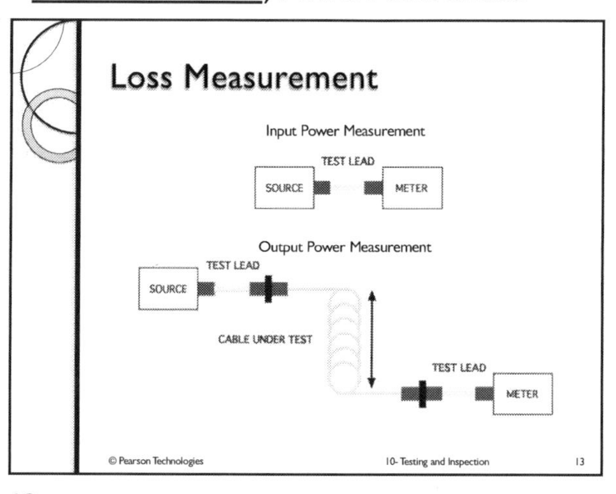

13

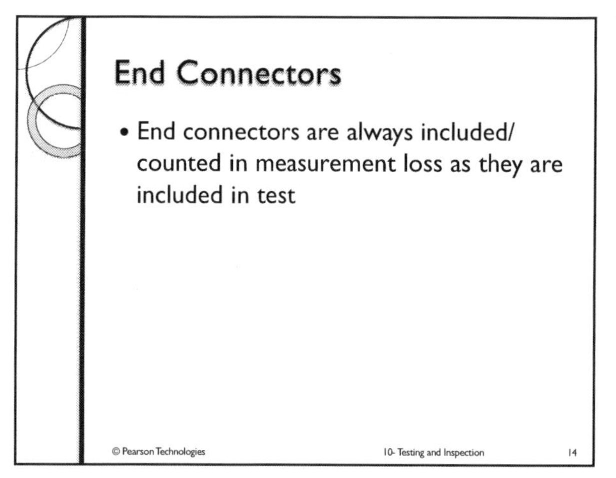

14

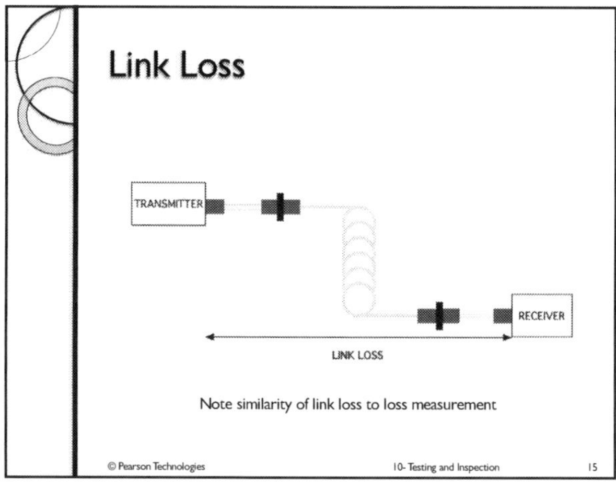

15

Part 1: Insertion Loss Testing

- Goals
- Insertion Loss Testing
  - Singlemode
  - Multimode
  - Range Testing
  - Standards
  - Test at Two Wavelengths
  - Bi-Directional Testing
  - Equipment Requirements

© Pearson Technologies 10- Testing and Inspection 16

16

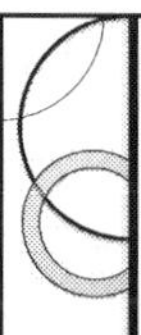

Multimode Testing

- Single lead input power measurement
  - Known as Method B of OFSTP-14 and EIA/TIA-526-14C
- Multimode loss depends on power distribution in core
- Current multimode, Gigabit sources create unique power distribution in core
  - Distribution known as 'encircled flux' (EF) and 'Restricted modal launch'

© Pearson Technologies 10- Testing and Inspection 17

17

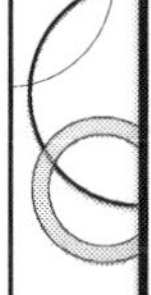

Multimode Transmitters

- Different distributions result in different power loss in link
- Two multimode transmitter active devices
  - LED
  - VCSEL requires restricted modal launch test method
- Distribution differences
  - LEDs used for 100 Mbps overfill core diameter and NA
  - VCSELs used for ≥ 1 Gbps underfill both

© Pearson Technologies 10- Testing and Inspection 18

18

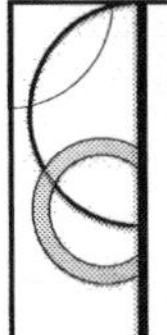

Key Understanding

- Testing with one type of active device does not predict loss with the other type

19

Multimode Test Procedures

- Multimode test standard is
  - ANSI-TIA-526-14-C
  - Also known as OFSTP-14
- Testing for both multimode active devices requires LED source
- Source wavelengths are 850nm and 1300nm
- Multimode testing has two test methods

20

Two Situations

- Situation 1: for ≥1 Gbps, use restricted modal launch test conditions
- Two test methods
  - Test with restricted modal launch source and standard test lead or
  - Test with LED source and restricted modal launch test lead at source
- Situation 2: test with LED source and test lead (100 Mbps) (uncommon)
  - With or without mandrel
  - (See Ch. 14, Professional Fiber Optic Installation, v10)

21

Short Version for Gigabit Transmission

- Test 50μ at 850nm with restricted modal launch conditions
  - Required by test standard
  - Restricted modal launch compliant source is an LED

22

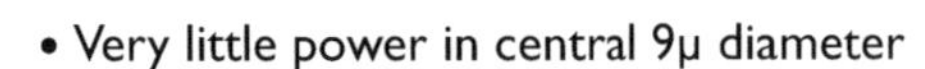

EF Launch Conditions

- Very little power in central 9μ diameter
- All power within 30μ diameter
  - Remember: core has 50μ diameter
- This compensates for non-typical RI profile in some early multimode fibers (See Ch. 8, Professional Fiber Optic Installation, v10)

23

Achieve Restricted Modal Launch Conditions with

- Restricted modal launch compliant source or
- Mode conditioner as source test lead at

24

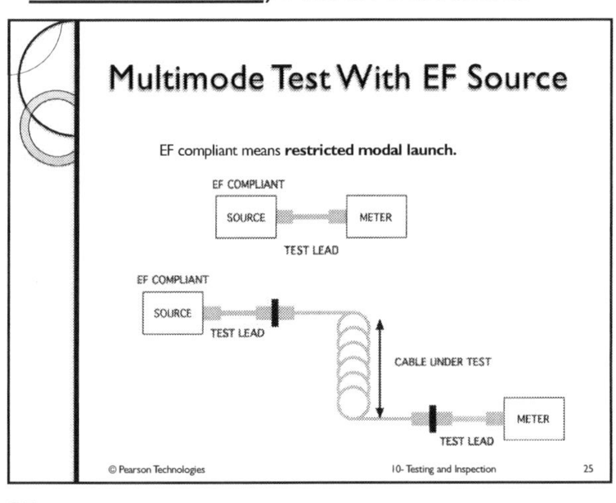

25

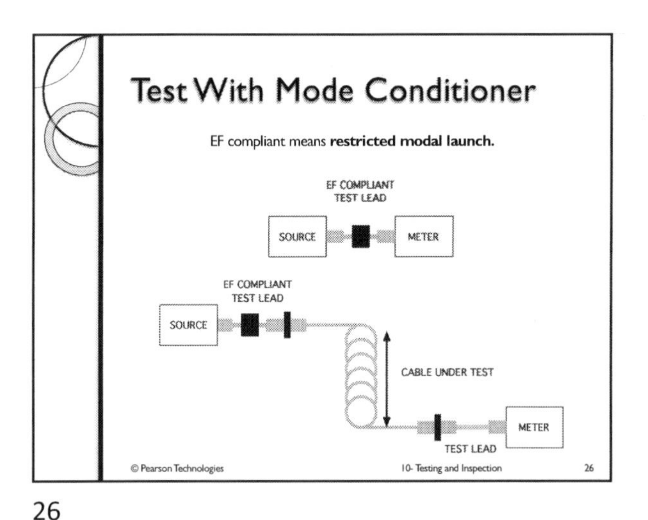

26

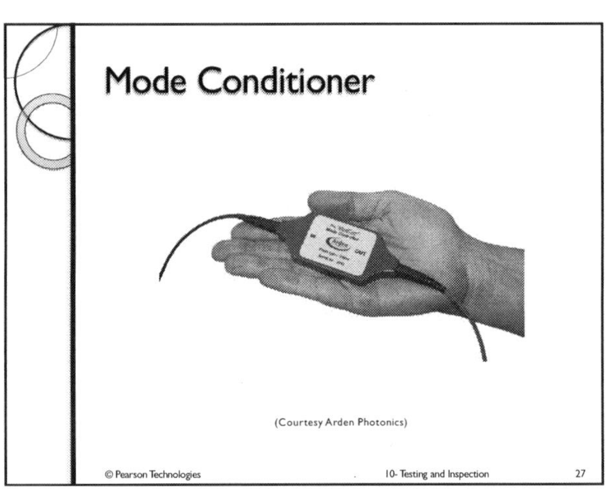

27

Perform Single Test/Fiber?

- Dirt can drop off or onto connectors during test
- TIA/EIA-568-D requires only one test
- Recommendation: perform two tests
  - If within 0.15 dB, accept value
  - If not, clean all connectors and retest
- This process prevents accepting values that include increased loss due to dirt

© Pearson Technologies 10- Testing and Inspection 28

28

Part 1: Insertion Loss Testing

- Goals
- Insertion Loss Testing
  - Singlemode
  - Multimode
  - Range Testing
  - Standards
  - Test at Two Wavelengths
  - Bi-Directional Testing
  - Equipment Requirements

© Pearson Technologies 10- Testing and Inspection 29

29

Perform A Single Insertion Loss Test?

- Will loss be same for each and every test?
  - Of course not
- If not, how will you interpret an increase in loss?
- Two interpretations possible for increase in loss:
  - Normal behavior or
  - Degradation of link components
- How to interpret increase?
  - Must know normal behavior
- How to define normal behavior?
  - Make multiple tests of multiple fibers

© Pearson Technologies 10- Testing and Inspection 30

30

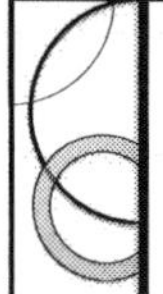

Recommendation: Test Range

- Range measurements may be made during
  - Initial installation
  - Troubleshooting or
  - As part of periodic maintenance testing

© Pearson Technologies 10- Testing and Inspection 31

31

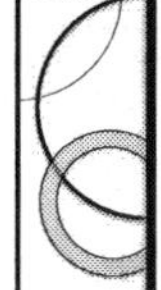

Determine Range 1

- Make 4-6 measurements of a link
- Calculate the difference between maximum and minimum losses
  - This is 'link range'
- Repeat this test on 4-6 links/fibers
- Calculate the difference between maximum and minimum losses for each link/fiber

© Pearson Technologies 10- Testing and Inspection 32

32

Determine Range 2

- Examine the 4-6 'link ranges'
- Choose the largest of the 4-6 ranges
- If they are ≥0.2 dB, chose the largest of the range values as the network range
- If largest range is ≤ 0.2 dB, use 0.2 dB as range
- Use network range to interpret increases in loss from future measurements

© Pearson Technologies 10- Testing and Inspection 33

33

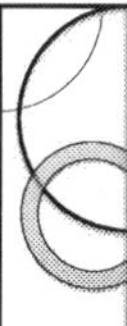

Use Of Range

- If the increase in loss between measurements made at different times is less than the range, the link exhibits no evidence of degradation.
- If this increase exceeds the range, degradation of link components has occurred.
  - Additional troubleshooting is required.

© Pearson Technologies 10- Testing and Inspection 34

34

Calculate Range

- If range is unknown, use 0.4 dB as range
- This value comes from repeatability of connector loss
- Repeatability is ≤0.2 dB/connector
- Range testing involves two end connectors, so twice the repeatability is the range
- Note: most range measurements are ~0.2 dB or less
- Determining range instead of calculating range enables earlier detection of degradation than the 0.4 dB value (opinion)

© Pearson Technologies 10- Testing and Inspection 35

35

Part 1: Insertion Loss Testing

- Goals
- Insertion Loss Testing
  - Singlemode
  - Multimode
  - Range Testing
  - Standards
  - Test at Two Wavelengths
  - Bi-Directional Testing
  - Equipment Requirements

© Pearson Technologies 10- Testing and Inspection 36

36

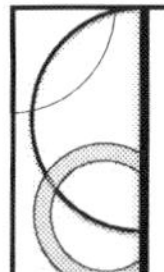

Methods Comply With Standards

- TIA/EIA-568-E
- TIA/EIA-526-14-C, aka OFSTP-14-A
- TIA/EIA-526-7-A, Edition 2, aka OFSTP-7
- IEC 61280-4-1
  - International standard

© Pearson Technologies 10- Testing and Inspection 37

37

TIA/EIA-568-E References

- Singlemode insertion loss testing
  - Method A.1 of TIA/EIA-526-7-A
- Multimode insertion loss testing
  - Method B of TIA/EIA-526-14-B
  - Source meets requirements of ANSI/TIA-455-78-B, aka 'Category 1' source
    - Category 1 source overfills core diameter and NA
- TIA/EIA-526-14-B references IEC 61280-4-1 for restricted modal launch testing

© Pearson Technologies 10- Testing and Inspection 38

38

Part 1: Insertion Loss Testing

- Goals
- Insertion Loss Testing
  - Singlemode
  - Multimode
  - Range Testing
  - Standards
  - Test at Two Wavelengths
  - Bi-Directional Testing
  - Equipment Requirements

© Pearson Technologies 10- Testing and Inspection 39

39

Multiple Wavelength Tests

- TIA/EIA-568-E requires testing at two wavelengths
  - 850nm and 1300nm
  - Exception: short horizontal links
  - On short horizontal links, loss difference due to wavelength difference deemed insignificant

© Pearson Technologies 10- Testing and Inspection 40

40

Recommendation

- Even if testing is not required at both wavelengths, test at both
- Reason: conditions of reduced reliability may not be evident at short wavelength but will at long wavelength
- Stress on fiber is a condition of reduced reliability

© Pearson Technologies 10- Testing and Inspection 41

41

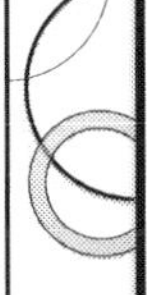

Stress Loss Vs. Wavelength

- Key understanding: as wavelength increases, sensitivity of fiber to stress increases
- Result: stress may not be evident at short wavelength but will be at long wavelength
- Consequence: to verify stress free installation, test at both wavelengths even if not required for other reason
- Examples

© Pearson Technologies 10- Testing and Inspection 42

42

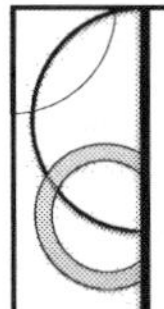

Part 1: Insertion Loss Testing

- Goals
- Insertion Loss Testing
  - Singlemode
  - Multimode
  - Range Testing
  - Standards
  - Test at Two Wavelengths
  - Bi-Directional Testing
  - Equipment Requirements

43

Bi-Directional Testing (14-3)

- Would you expect that the insertion loss is *exactly* the same in both directions?
- Of course not

44

Bi-Directional Testing

- TIA/EIA-568-D allows testing in one direction
- Directional effects exist, but are small
- Differences caused by
  - Core diameter differences
  - NA differences
  - Differential modal attenuation
  - Core offset
  - Cladding Ovality
  - Fiber offset in connectors

45

Part 1: Insertion Loss Testing

- Goals
- Insertion Loss Testing
  - Singlemode
  - Multimode
  - Range Testing
  - Standards
  - Test at Two Wavelengths
  - Bi-Directional Testing
  - Equipment Requirements

46

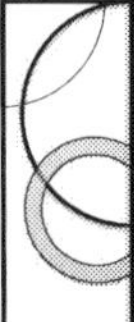

Equipment Requirements (14-9)

- Source and meter known as optical loss test set (OLTS)
- Stabilized light source
  - Multimode matched wavelengths: 850nm, 1300nm
    - LED for multimode testing
    - EF compliant source or EF mode conditioner
  - Singlemode matched wavelengths: 1310nm, 1550nm
    - Laser for singlemode testing
- Calibrated power meter
  - Meter with changeable adapters (opinion)
- Two qualified reference leads
- Low loss barrels

47

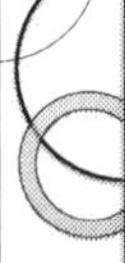

Light Source

- Stabilized = constant output power
  - Make many measurements without checking input power level
- Wavelength is same as that of transmitter
- Launch conditions
  - Restricted modal launch compliant
    - Required for testing 50μ @ 850 nm
    - Can use for all testing
  - Category 1 source: for 100 Mbps, multimode testing

48

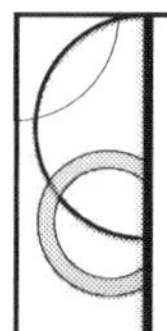

Power Meter

- Calibrated and traceable to NIST at all wavelengths
- Offset function
- Changeable adapters
  - May be required to perform multimode Method B
  - Not required by TIA/EIA-526-14-B

© Pearson Technologies 10- Testing and Inspection 49

49

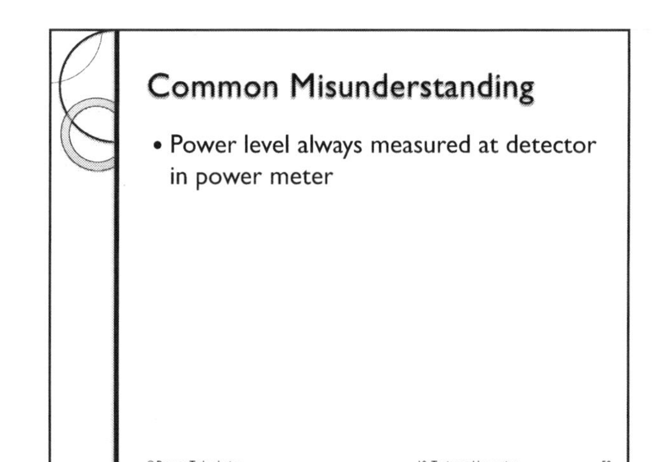

50

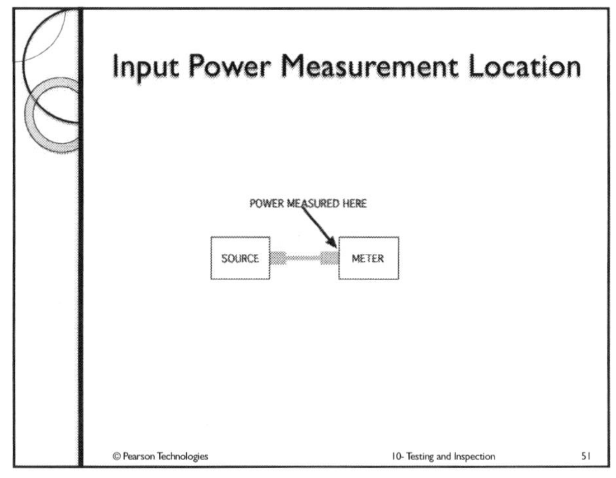

51

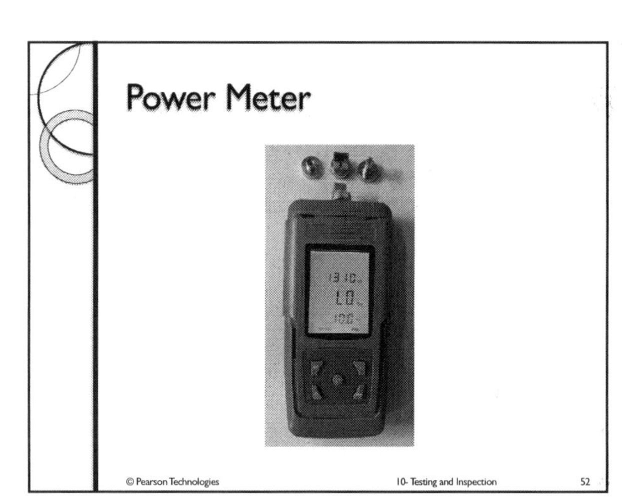

52

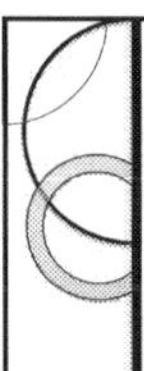

Reference Leads

- Be qualified
- Be 2-5 m long
- Have the same core diameter and connector type as on cables to be tested
  - Exception: testing connector types with same ferrule diameter with barrel accepting different connector types on each end

© Pearson Technologies 10- Testing and Inspection 53

53

Multimode Reference Lead Qualification

- TIA/EIA-568-E multimode method requires
  - Multimode reference lead with restricted modal launch
  - In other words, perform test lead qualification in manner same as test is to be performed

© Pearson Technologies 10- Testing and Inspection 54

54

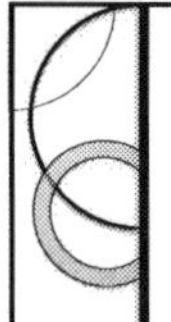

Qualification Acceptance Values

- Two Test Lead Grades
  - Standard
    - Qualified if ≤ 0.75 dB
  - Reference grade
    - Qualified if ≤ 0.1 dB (obsolete value)
    - Qualified if ≤ 0.5 dB (current value in ANSI-/TIA-568.3-E)

© Pearson Technologies 10- Testing and Inspection 55

55

Reference Grade Test Leads

- Use recommended but not required
- Have tighter tolerances on fiber and connectors
- More expensive than standard test leads
  - 8 times cost of standard test leads (7/22/23)

© Pearson Technologies 10- Testing and Inspection 56

56

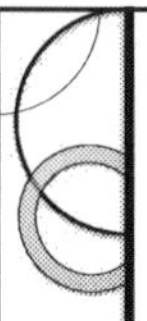

Recommendation (Opinion)

- Use as qualification acceptance value
  - 0.5 dB instead of 0.75 dB allowed by TIA/EIA-568-D
- Test leads should have loss lower than cables being tested
- Risk of using 0.75 dB: acceptance of high loss products by using test leads that are 'high' loss
- This author's experience is that most qualification measurements are < 0.4 dB

© Pearson Technologies 10- Testing and Inspection 57

57

Test Lead Qualification

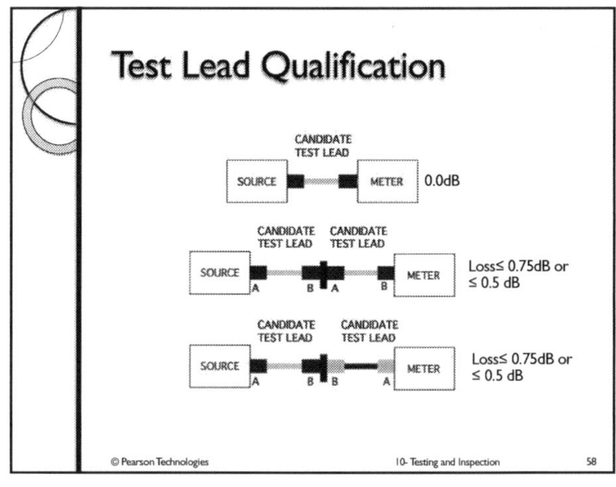

© Pearson Technologies 10- Testing and Inspection 58

58

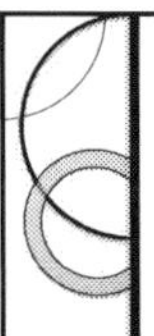

Lead Maintenance

- Prior to testing, the installer cleans and inspects all reference leads with a micro-scope.
- Cleaning can restore reference leads to low loss and low reflectance
- During testing, the installer cleans the connectors on test leads frequently to remove small particles of dirt and dust, which can cause high loss measurements.
  - Since cleaning can contaminate ferrule ends, technician performs periodic microscopic inspection of test leads

© Pearson Technologies 10- Testing and Inspection 59

59

Do Reference Leads Effect Measurement?

- You might argue that the loss of the reference lead will influence the measured loss
- You would be correct
- Variation from lead to lead is small, ~ 0.2 dB

© Pearson Technologies 10- Testing and Inspection 60

60

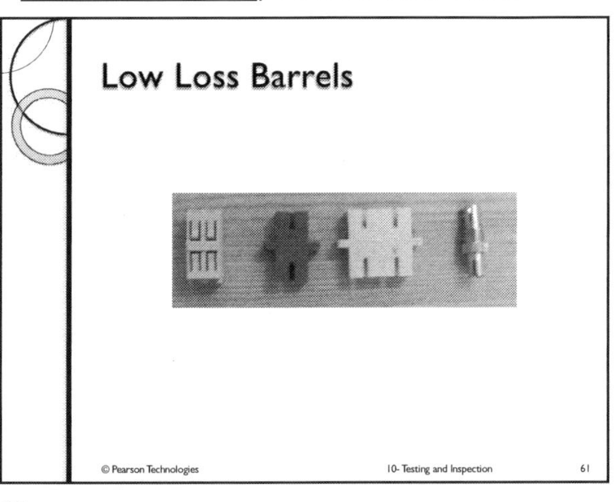

61

Barrels

- Have a precision alignment sleeve
- Sleeves are
  - Plastic
  - Copper beryllium
  - Ceramic
- Recommendations
  - Do not use barrels with plastic sleeves
  - Use use barrels with ceramic sleeves: they wear slowly and do not need frequent replacement

© Pearson Technologies 10- Testing and Inspection 62

62

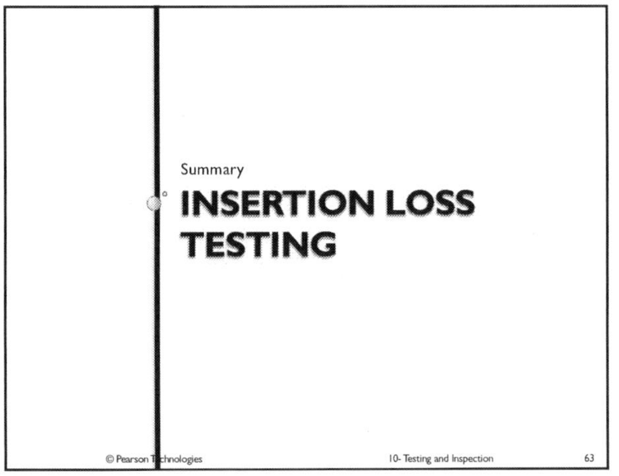

63

Summary: Singlemode Testing

- Input measurement
  - Single lead method
- Test source
  - Stabilized
  - Wavelength matches that of transmitter
- Power meter
  - Traceable calibration
  - Changeable adapters
- Test leads are
  - Matched
  - Qualified
  - 2-5 m long
- Barrels are low loss

© Pearson Technologies 10- Testing and Inspection 64

64

Summary: Multimode Testing

- Input Measurement
  - Is single lead input power measurement
  - Is EF compliant
- Test source
  - Stabilized
  - Wavelength matches that of transmitter
  - EF compliant or Category I source
- Power meter
  - Traceable calibration
  - Changeable adapters
- Test leads are
  - Matched
  - Qualified
  - 2-5 m long
- Barrels are low loss

© Pearson Technologies 10- Testing and Inspection 65

65

Hands-On Insertion Loss Testing

- Do not proceed until you hear the magic word: Proceed
- Multimode
  - With test lead, connect source to meter
  - Set reference to 0 dB
  - Remove test lead from meter
  - With barrel, attach source test lead to end of cable
  - With barrel, attach second test lead to opposite end of cable
  - Read and record loss

© Pearson Technologies 10- Testing and Inspection 66

66

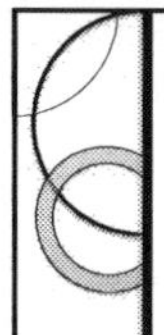

Hands-On Range Testing

- Do not proceed until you hear the magic word: Proceed
- Multimode
  - Perform insertion loss test
  - Record loss
  - Disconnect and reconnect both ends
  - Record loss
  - Repeat for total of 6 measurements
  - Calculate range= maximum- minimum

67

Hands-On Insertion Loss Testing

- Do not proceed until you hear the magic word: Proceed
- Singlemode
  - With test lead, connect source to meter
  - Set reference to 0 dB
  - Remove test lead from meter
  - With barrel, attach source test lead to end of cable
  - With barrel, attach second test lead to opposite end of cable
  - Read and record loss

68

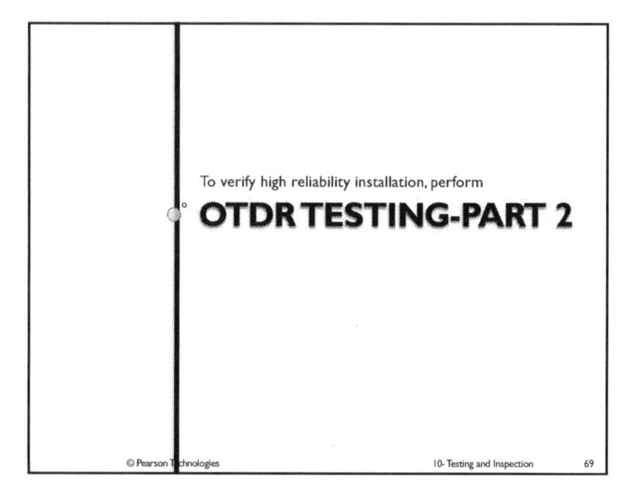

To verify high reliability installation, perform

OTDR TESTING-PART 2

69

Part 2: OTDR Testing

- General Information
- Purpose
- Types
- Principles of Operation
- Information Produced
- Three Basic Traces
- Using the OTDR
- Manual Measurements

70

OTDR Test Standards

- FOTP-61 for attenuation rate testing
  - also known as EIA/TIA-455-61
- FOTP-8, Measurement of Splice or Connector Loss and Reflectance Using an OTDR
- FOTP-59 Measurement of Fiber Point Defects Using an OTDR
- FOTP-133-A IEC 60793-1-22 Measurement Methods and Test Procedures - Length Measurement

71

Insertion Loss Test Disadvantages

- Test is blind to location and distribution of loss
- Conditions of reduced reliability can result in acceptable insertion loss
- Key understanding: low loss link components can compensate for high loss components

72

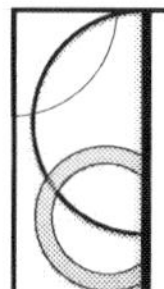

Additional Test Needed

- We need an additional test!
- To indicate that all components in a link are properly and reliably installed
- Components are
  - Connectors
  - Cable segments
  - Splices
- That Test Is…

73

Optical Time Domain Reflectometry

PS: It's ok to say OTDR

See Ch. 15, Professional Fiber Optic Installation, v10

74

OTDR Testing, Part 2

- General Information
- Purpose
- Types
- Principles of Operation
- Information Produced
- Three Basic Traces
- Using the OTDR
- Manual Measurements

75

Purpose 

- To verify that the power loss in each component in a link is proper
- A component is a
  - Cable
  - Connector
  - Splice
  - Passive device

76

OTDR

- OTDR testing does not replace insertion loss testing, as it does not simulate link operation.
- Enables testing of almost every element in a link (See 'dead zone')
- Explicitly indicates reliability
- Surprise: only one method!
- OTDR measures
  - Fiber attenuation rate
  - Splice and connector loss
  - Segment lengths
- OTDR enables finding faults; i.e., breaks and locations of excessive loss

77

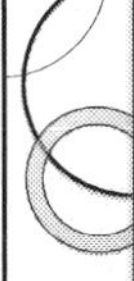

Three OTDR Tests

- OTDR testing can be performed
  - On cable as received or before installation
  - After cable installation
  - After splicing
  - After all installation activities

78

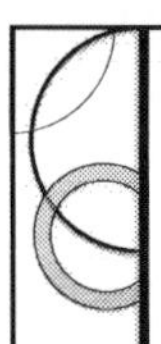

Part 2: OTDR Testing

- General Information
- Purpose
- Types
- Principles of Operation
- Information Produced
- Three Basic Traces
- Using the OTDR
- Manual Measurements

© Pearson Technologies 10- Testing and Inspection 79

79

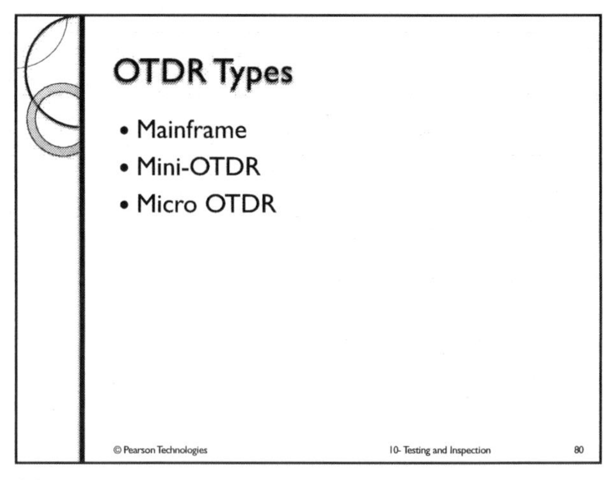

80

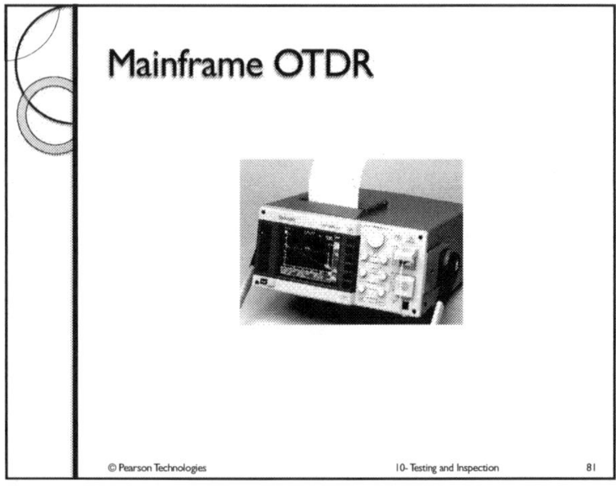

81

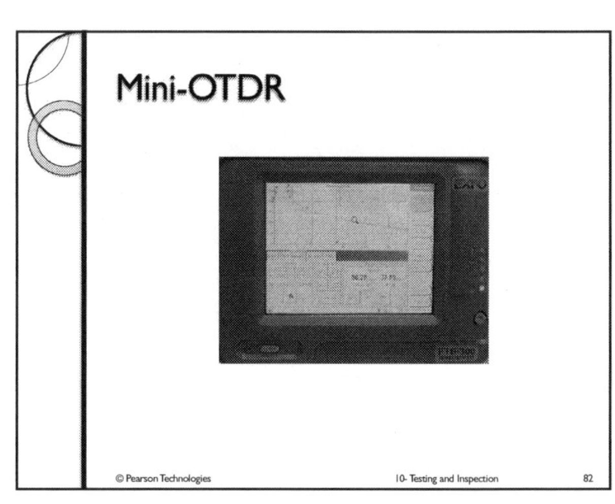

82

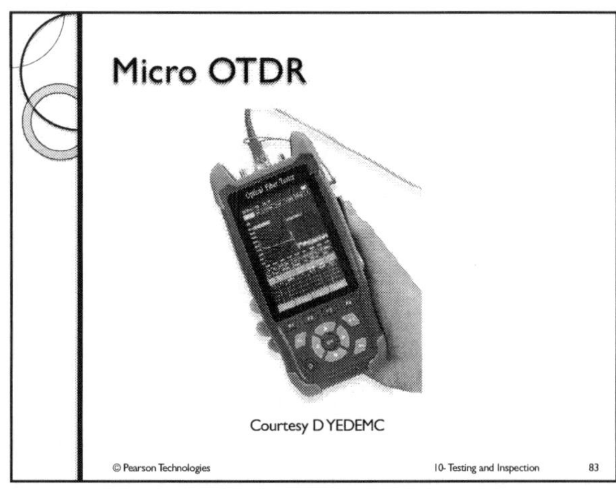

83

Part 2: OTDR Testing

- General Information
- Purpose
- Types
- Principles of Operation
- Information Produced
- Three Basic Traces
- Using the OTDR
- Manual Measurements

© Pearson Technologies 10- Testing and Inspection 84

84

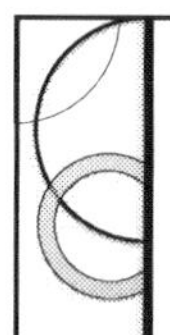

OTDR Principles

- Power loss (attenuation) results from scattering of light by atoms
- This mechanism is called Rayleigh scattering (Chapter 3)
- Lost power is scattered towards core-cladding boundary outside angle defined by NA
- Very low power level is scattered backwards towards boundary inside angle defined by NA
  - The OTDR measures differences in backscattered power levels
- Some light is always traveling backwards in the fiber!
- OTDR operates in a manner analogous to radar

© Pearson Technologies 10- Testing and Inspection 85

85

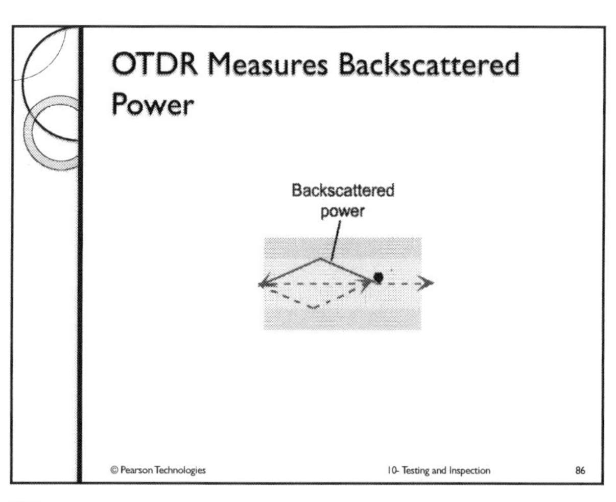

86

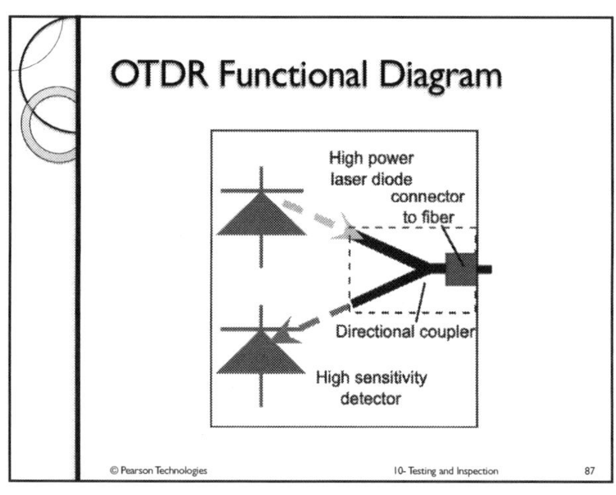

87

Theoretical OTDR Trace

Log(power)

0 Distance

© Pearson Technologies 10- Testing and Inspection 88

88

Multiple Measurements

- Backscattered power is low
- Amplification of low power by OTDR adds noise
- OTDR makes multiple tests of link, displaying averaged power levels
  - My experience: 10,000-100,000 tests
- Averaging reduces noise to level that enables accurate measurements

© Pearson Technologies 10- Testing and Inspection 89

89

OTDR Displays Power Differences

- Measure attenuation rate from difference of power levels at beginning and end of segment
- Measure connection loss from difference of power levels before and after connection
- Can measure power loss of almost every component in link

© Pearson Technologies 10- Testing and Inspection 90

90

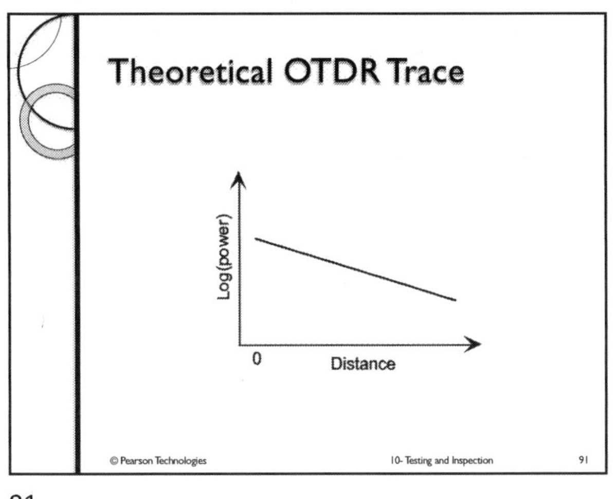

91

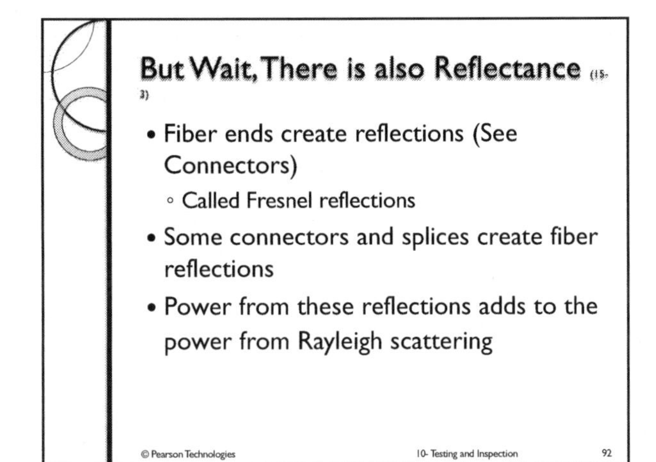

92

Add Fresnel Reflection Power

Power level 2
Power level 3
Power level 1
Log(power)
0
Distance

© Pearson Technologies 10- Testing and Inspection 93

93

Remember

- Horizontal axis is time
- The OTDR measures round trip travel time
- The OTDR calculates distance from the RI
- In the previous slide, 0 distance for reflections distance means 0 time

© Pearson Technologies 10- Testing and Inspection 94

94

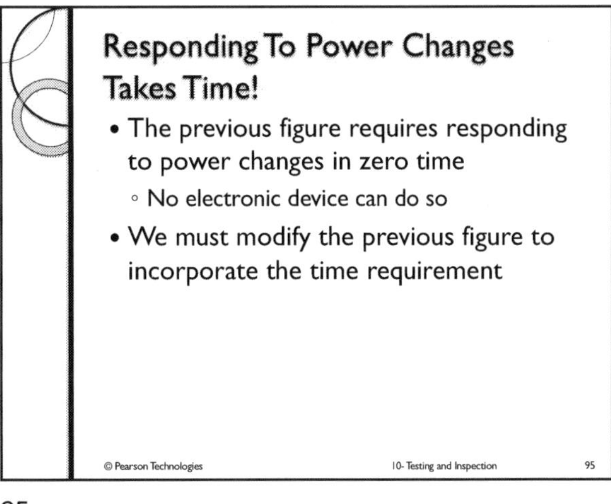

95

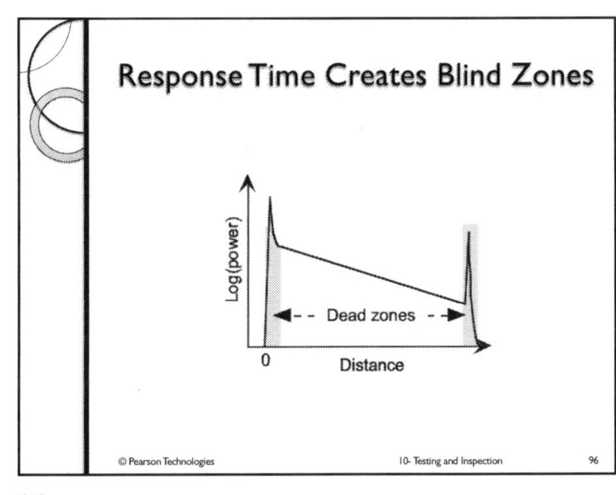

96

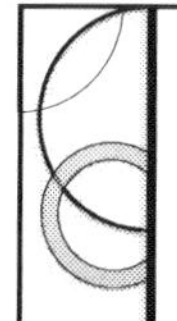

'Blind Zones', aka 'Dead Zones'

- Are lengths in which you cannot measure loss of individual features (connectors, splices, etc.)
- The loss of features more closely spaced than the width of the dead zone cannot be measured
  - Key point: we cannot measure the loss of *all* components in a link
  - Key point: the dead zone is created by the OTDR, not by features in the link
- The dead zone is the reason that use of an OTDR on indoor cables may be of limited value
  - Due to the limited resolution created by the dead zone

© Pearson Technologies 10- Testing and Inspection 97

97

Concealed Features

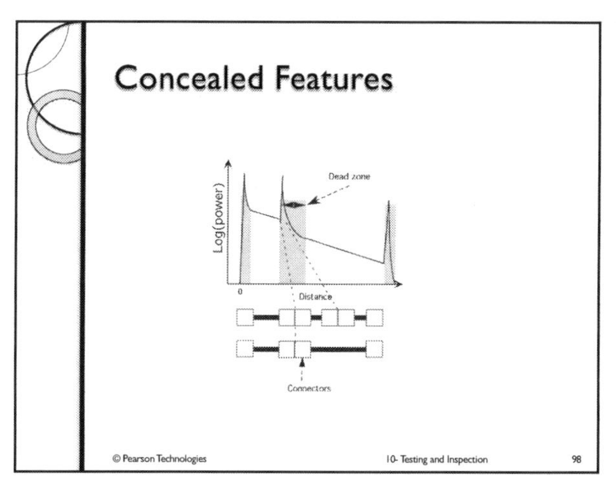

© Pearson Technologies 10- Testing and Inspection 98

98

Most Important Rule

- For interpretation of OTDR traces
  - You cannot build a map from a trace
    - Dead zones may conceal features!
  - You can interpret a trace properly only with a map

© Pearson Technologies 10- Testing and Inspection 99

99

OTDR Testing, Part 2

- General Information
- Purpose
- Types
- Principles of Operation
- Information Produced
- Three Basic Traces
- Using the OTDR
- Manual Measurements

© Pearson Technologies 10- Testing and Inspection 100

100

Three Basic Traces (15-5)

- Reflective loss
- Non-reflective loss
- Bad launch

© Pearson Technologies 10- Testing and Inspection 101

101

Basic Trace #1: Reflective Loss

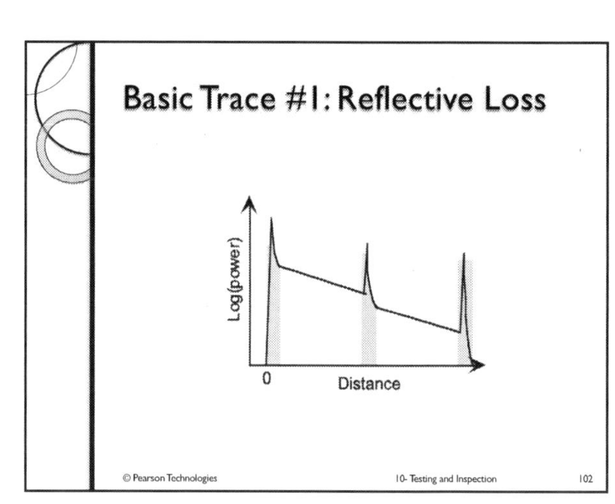

© Pearson Technologies 10- Testing and Inspection 102

102

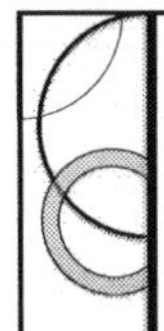

..From At Least Five Configurations

- Two segments connected by radius connectors
- Two multimode segments connected by mechanical splice
- Two singlemode segments connected by a mechanical splice
- A broken fiber and a tight tube cable
- A single cable segment that is exhibiting multiple reflections, also known as ghost reflections

© Pearson Technologies 10- Testing and Inspection 103

103

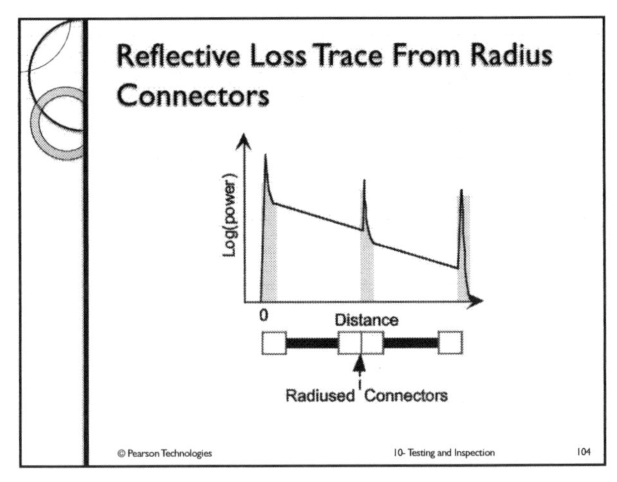

104

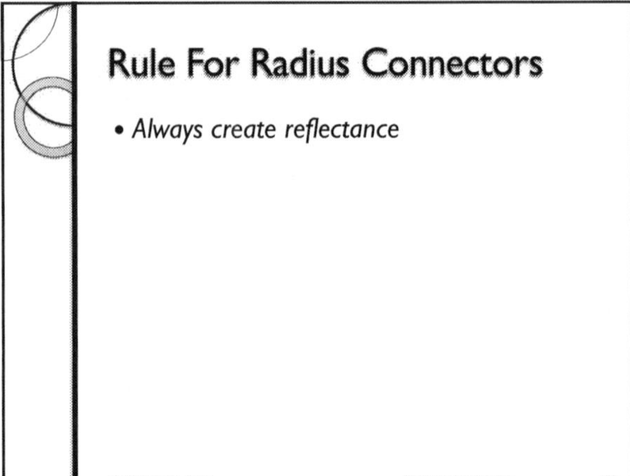

105

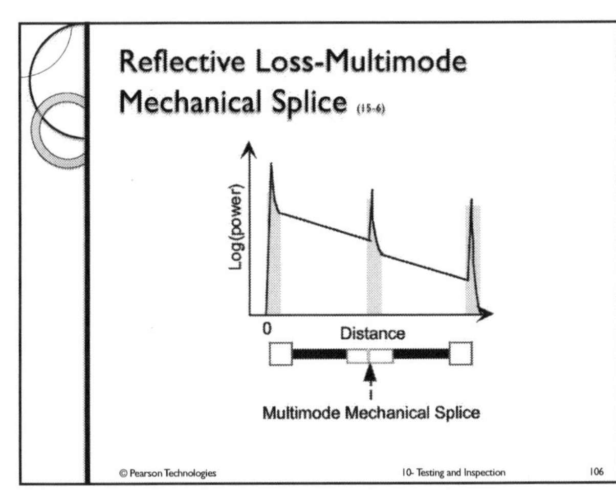

106

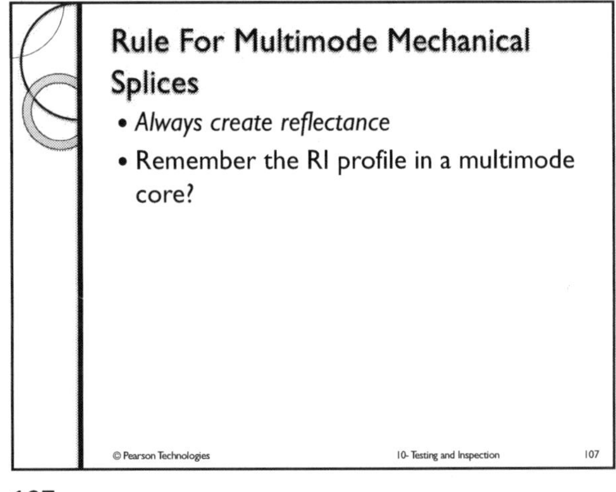

107

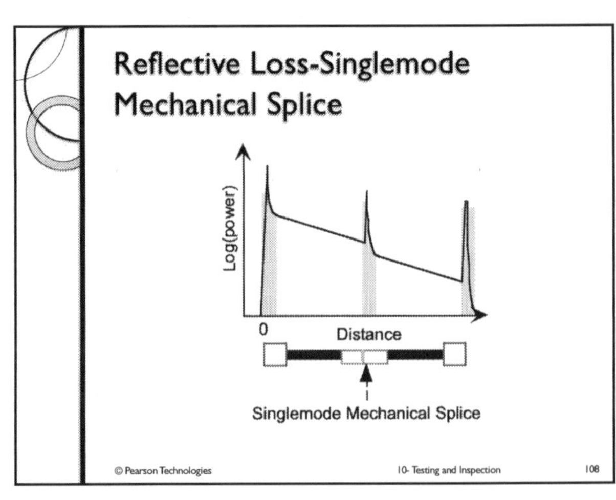

108

Rule For Singlemode Mechanical Splices

- *May* create reflectance
- Remember: reflectance is created by a change in RI
- If the RIs of both fibers and that of the index matching gel are not the same, there is reflectance

© Pearson Technologies 10- Testing and Inspection 109

109

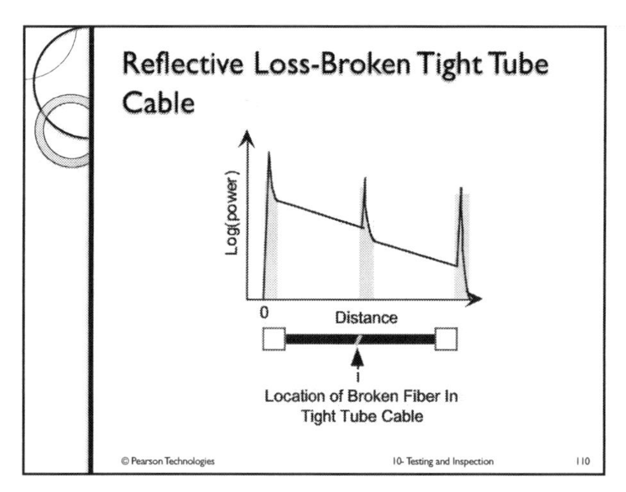

110

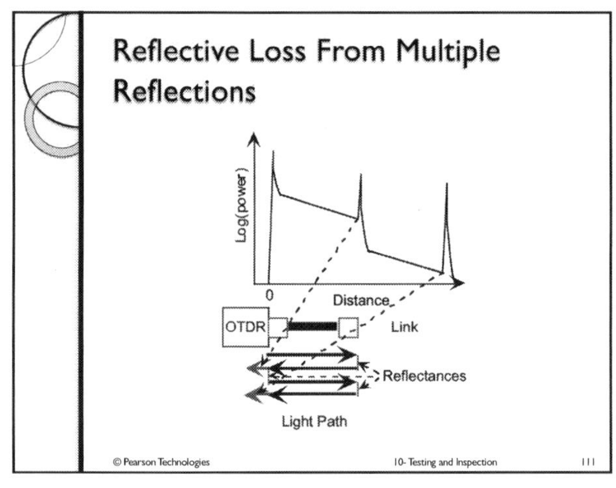

111

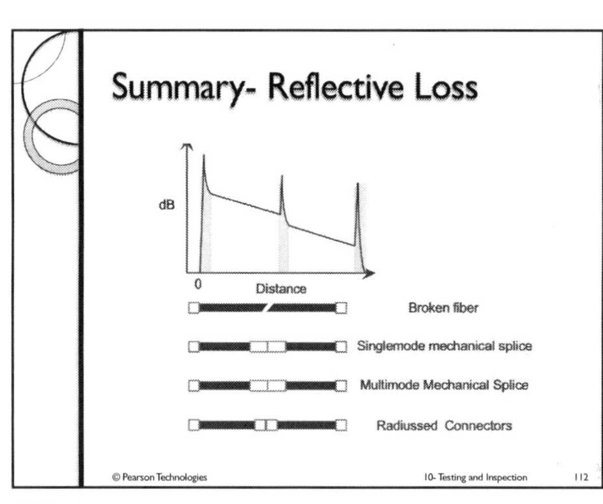

112

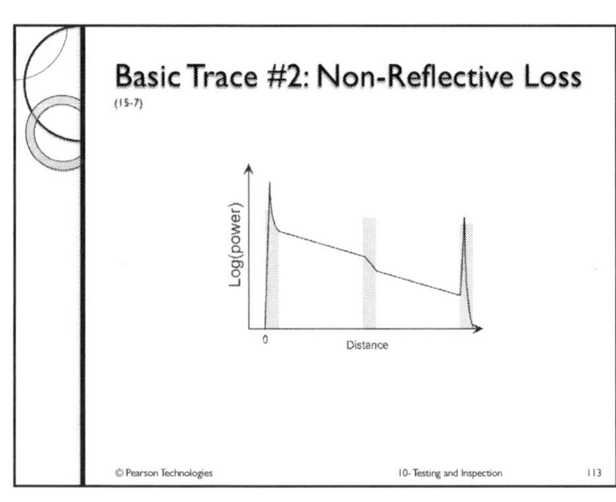

113

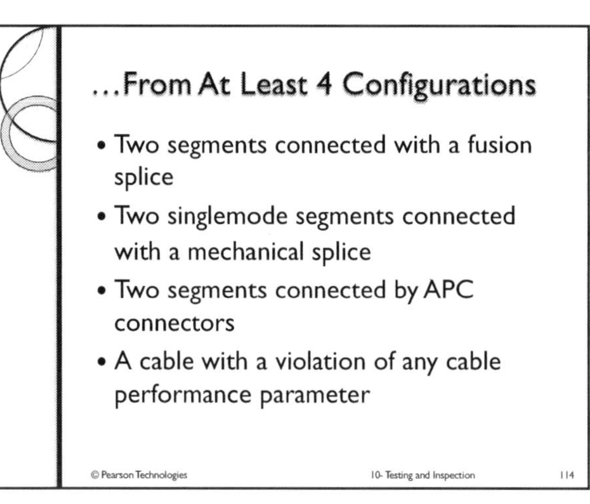

114

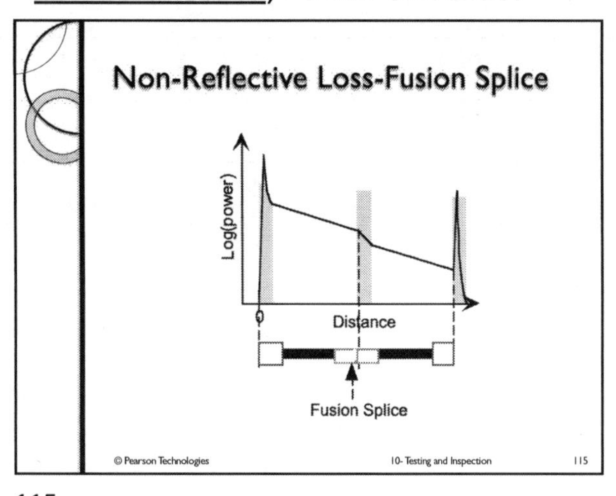

115

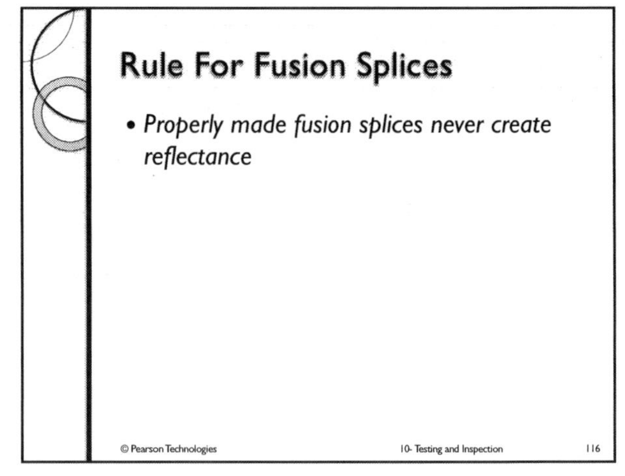

116

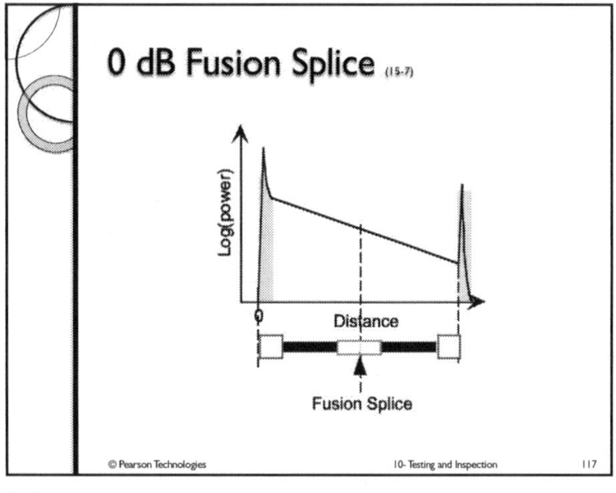

117

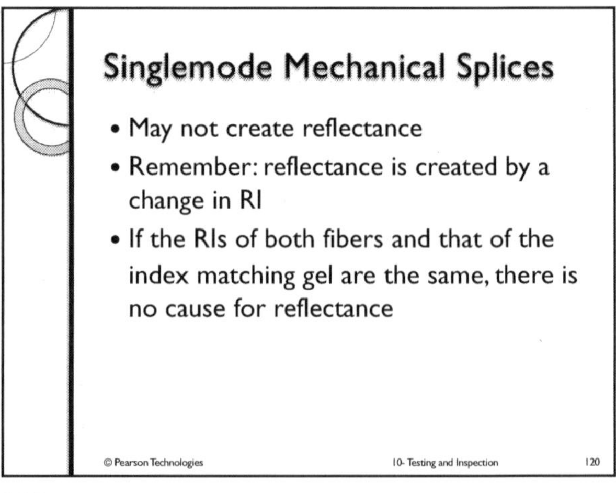

118

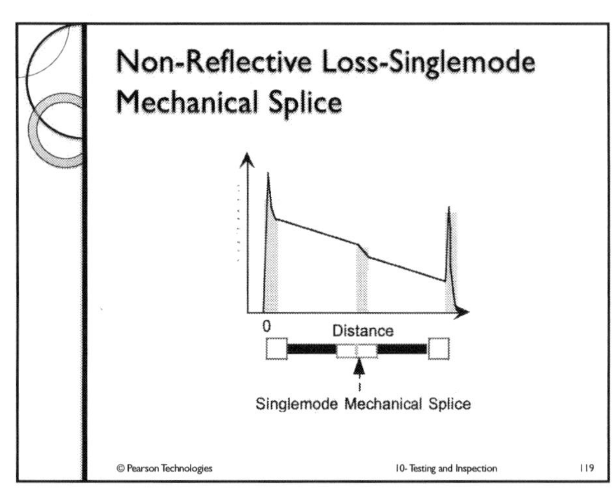

119

120

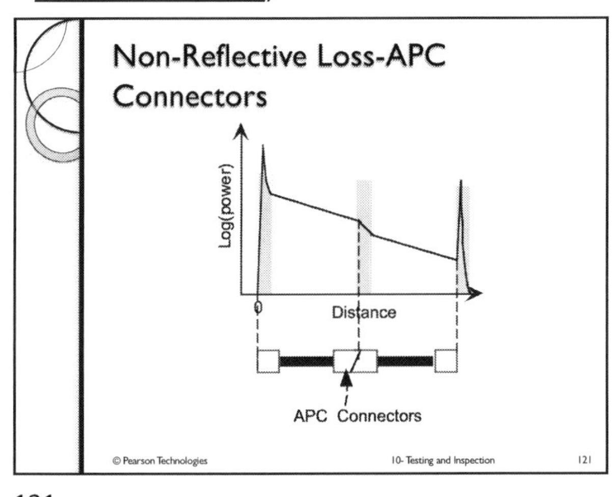

121

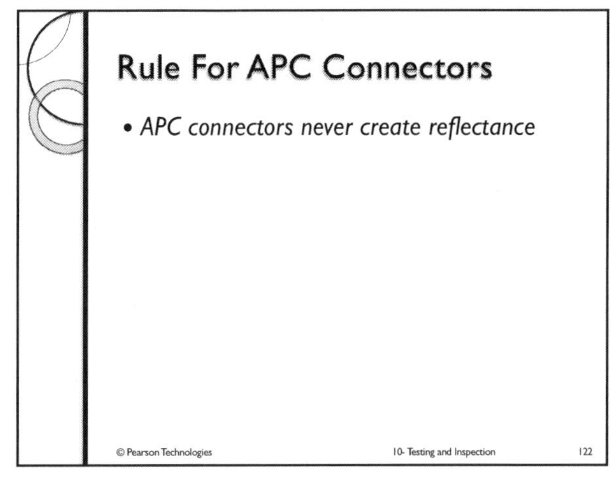

122

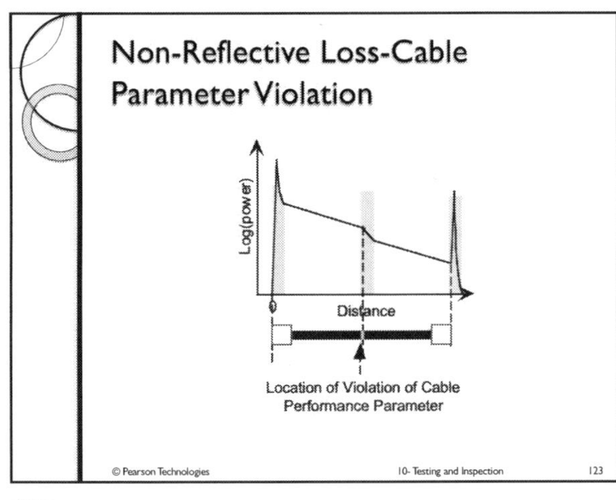

123

Summary- Non-Reflective Loss

dB

0

Distance

Fusion splice

Fusion splice

Singlemode mechanical splice

APC connectors

Violation of cable performance parameter

© Pearson Technologies 10- Testing and Inspection 124

124

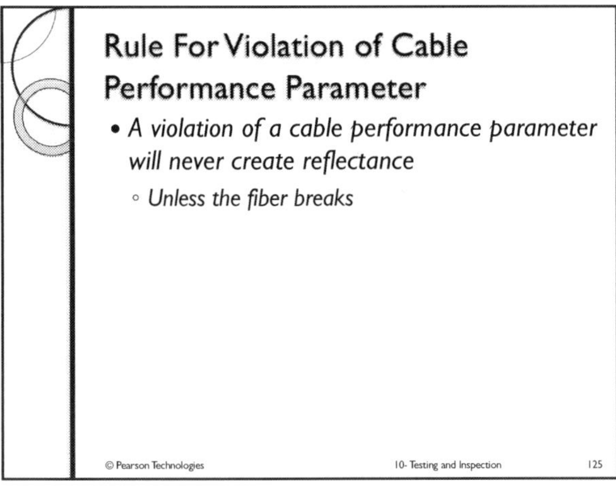

125

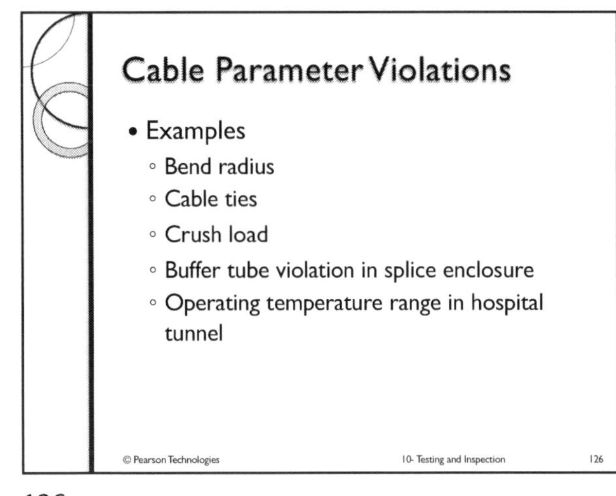

126

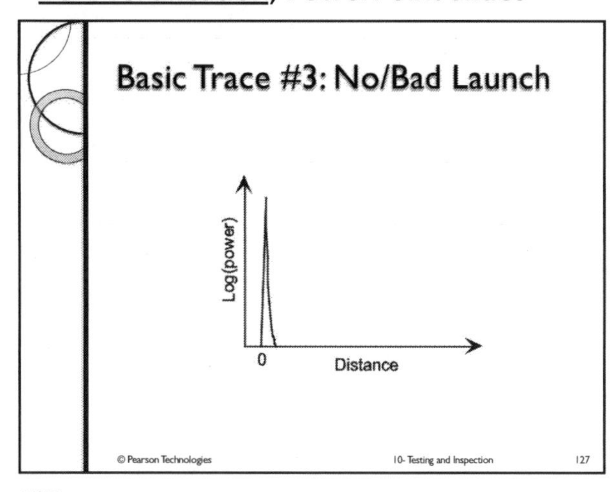

127

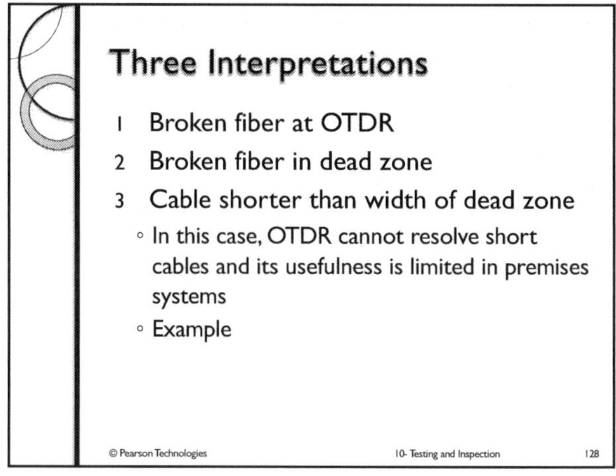

128

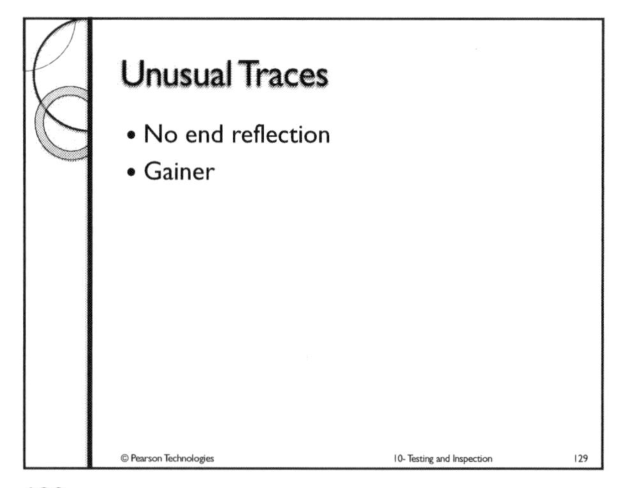

129

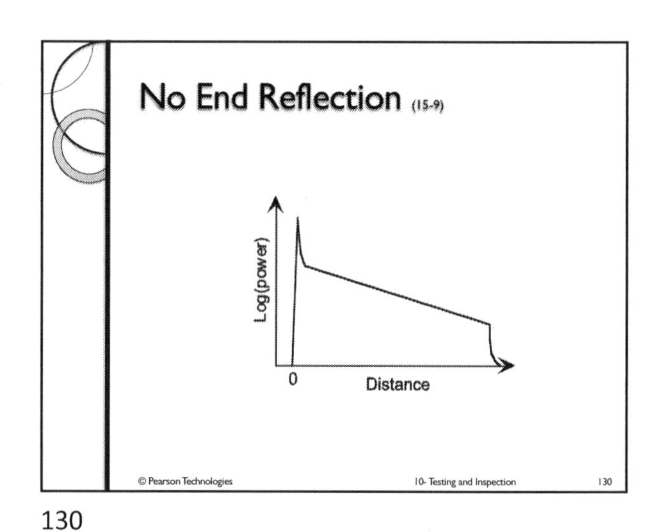

130

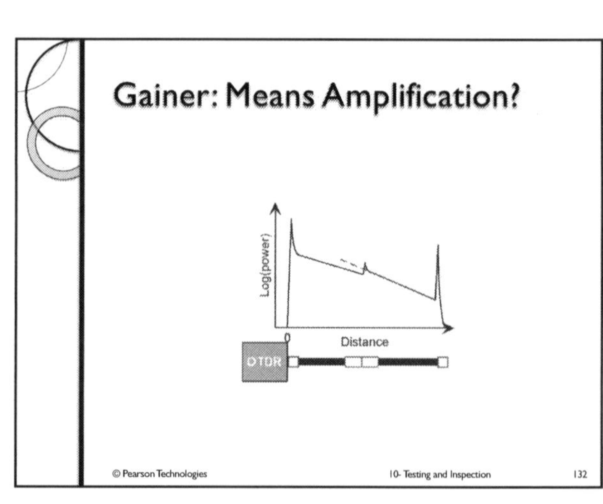

131

Gainer: Means Amplification?

Log(power)

0

Distance

OTDR

© Pearson Technologies 10- Testing and Inspection 132

132

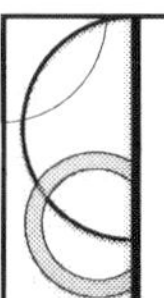

No!

- Gainers result from differences in fiber
  - Attenuation rates
  - Core diameters
  - MFDs

© Pearson Technologies 10- Testing and Inspection 133

133

Cause Of Gainers

- Multimode gainers
  - Small core to large core
  - Low attenuation rate to high attenuation rate
- Singlemode gainers
  - Large MFD to small MFD
  - Low attenuation rate to high attenuation rate

© Pearson Technologies 10- Testing and Inspection 134

134

Consequence

- Fibers can bias splice loss up in one direction and down in opposite direction
- Important Rule:
  - True splice loss is average of loss in both directions
  - Example

© Pearson Technologies 10- Testing and Inspection 135

135

Bi-Directional Testing

- Must be done to verify true splice loss values (See Testing Part 2, OTDR)

© Pearson Technologies 10- Testing and Inspection 136

136

Multiple Reflections

- Caused by light returning to OTDR that is reflected back into fiber for additional round trips
- Location of ghost reflection is a multiple of segment length creating ghost

© Pearson Technologies 10- Testing and Inspection 137

137

Ghost Origin

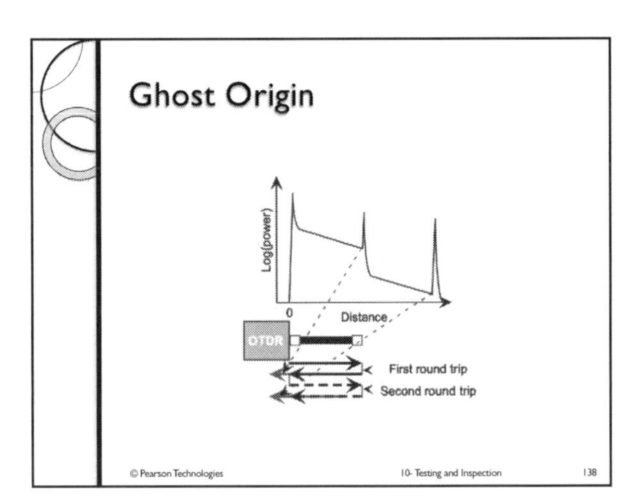

© Pearson Technologies 10- Testing and Inspection 138

138

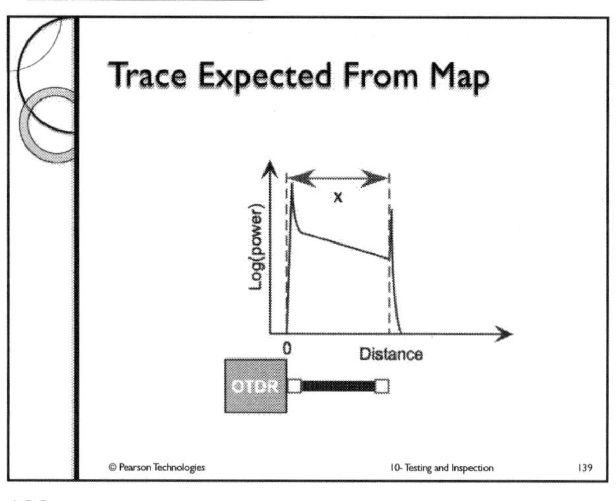

139

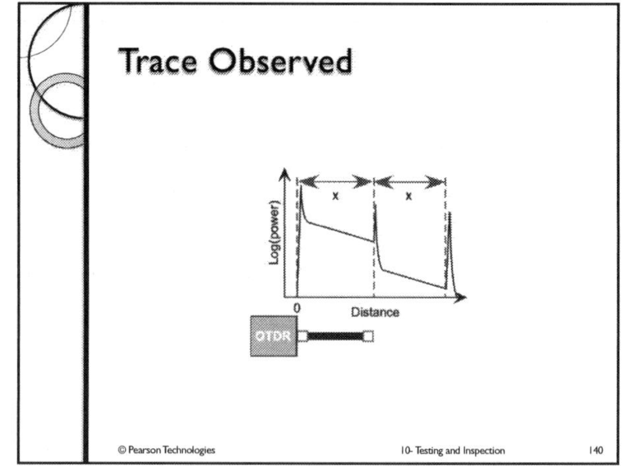

140

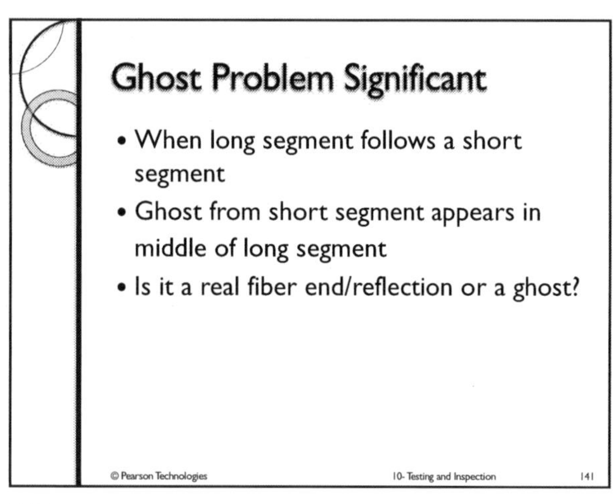

141

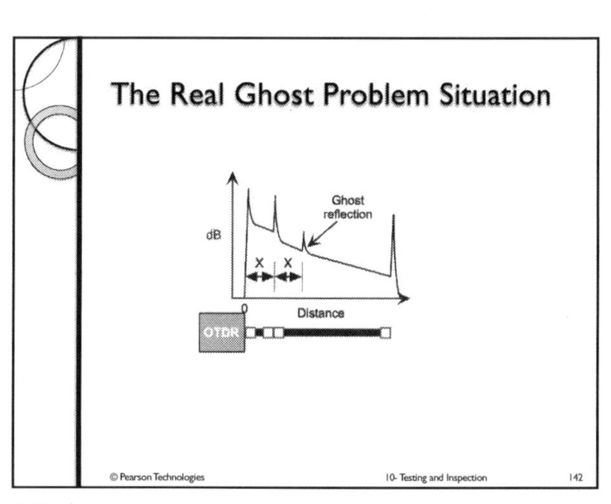

142

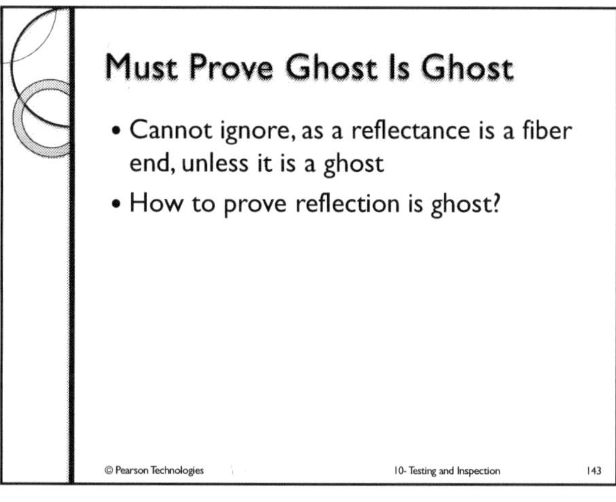

143

The Solution

- Test in both directions
- If peak does not appear in both directions, it cannot be real
- If peak appears at same location in both tests, it cannot be real
- If peak moves to appropriate location in test from second end, it is real

© Pearson Technologies 10- Testing and Inspection 144

144

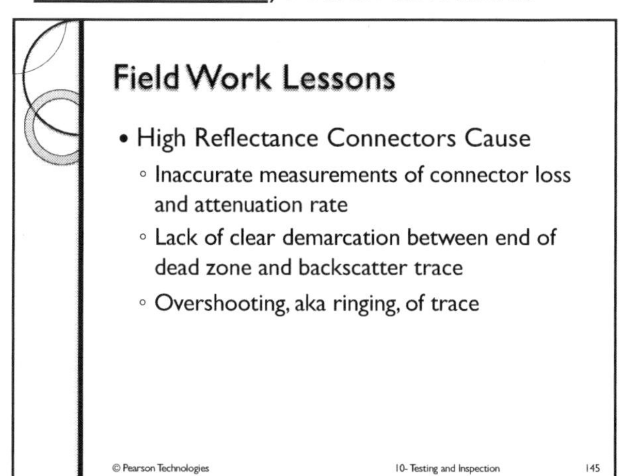

145

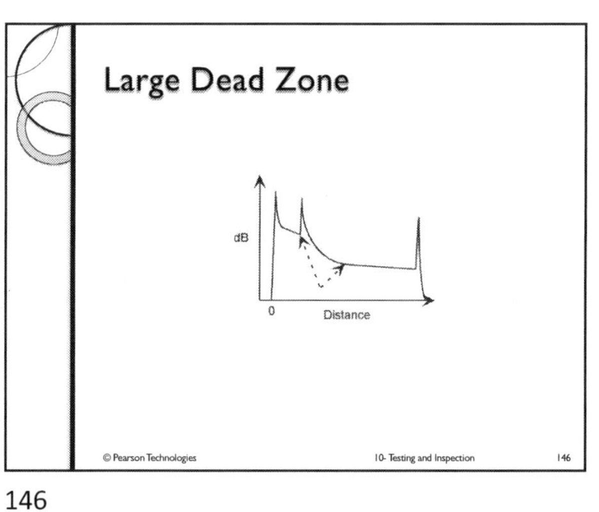

146

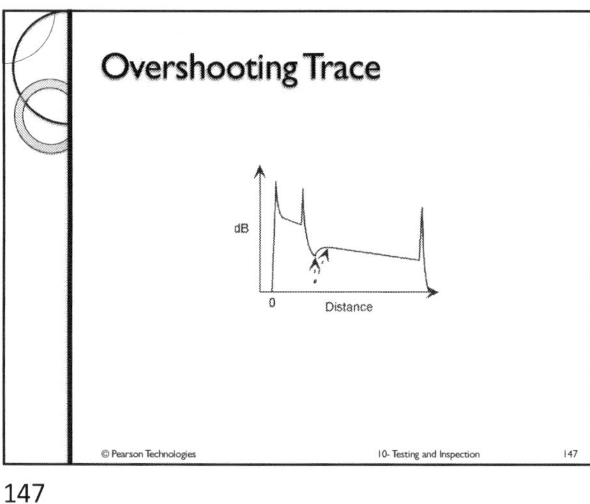

147

Wavelength Effect

- Same as in insertion loss tests
  - Increased sensitivity to stress at increased wavelength
- OTDR test at two wavelengths can differentiate between
  - High loss splice that requires replacement
  - Cable or buffer tube routing problem that can be fixed easily

148

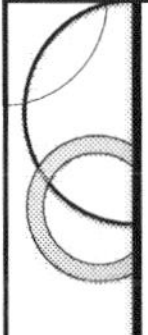

OTDR Testing, Part 2

- General Information
- Purpose
- Types
- Principles of Operation
- Information Produced
- Three Basic Traces
- Using the OTDR
  - Setting Up
  - Manual measurements

149

Data Required For Set Up(15-12)

- Wavelength of operation
- Pulse width
- Maximum length of cable to be tested
- Index of refraction and
- Maximum time allowed for the test or the number of pulses to be analyzed by the OTDR
- Backscatter coefficient
  - Calibrates OTDR to fiber under test

150

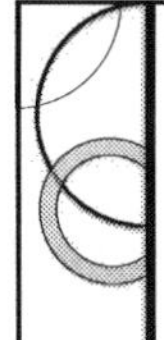

Wavelength

- Same as that of optoelectronics

© Pearson Technologies 10- Testing and Inspection 151

151

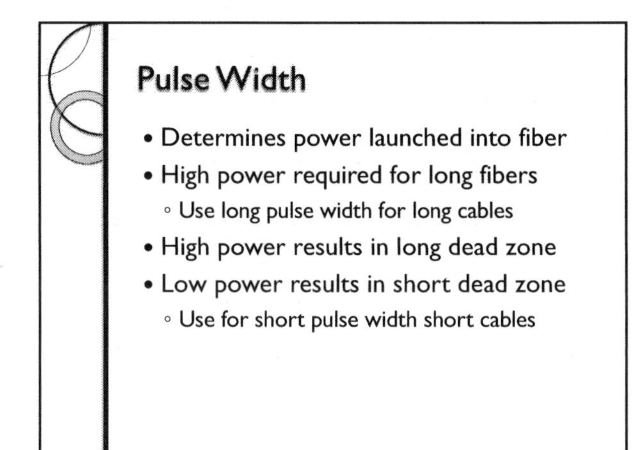

152

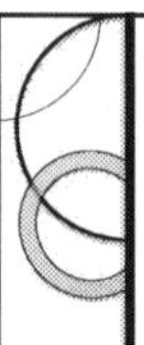

Maximum Length Of Cable

- Determines testing time
- Set to value longer than, but close to, length of cable to be tested

© Pearson Technologies 10- Testing and Inspection 153

153

Refractive Index

- Of fiber under test
- Enables accurate length and attenuation rate measurements

© Pearson Technologies 10- Testing and Inspection 154

154

Typical RI Values

Product		850 nm	1300 nm	1310 nm	1550 nm
Draka MaxCap OM-2	50/125	1.482	1.477		
Draka MaxCap OM-3	50/125	1.482	1.477		
Draka MaxCap OM-4	50/125	1.482	1.477		
Laserwave 550-300	50/125	1.483	1.479		
LaserWave G+	50/125	1.483	1.479		
InfiniCor 50	50/125	1.481	1.476		
InfiniCor 62.5	62.5/125	1.496	1.491		
SMF-28e+				1.4676	1.4682
Draka singlemode G.652				1.4670	1.4680
OFS AllWave ZWP				1.4670	1.4680

© Pearson Technologies 10- Testing and Inspection 155

155

Maximum Time For Test

- Determines accuracy of measurements by reducing noise in trace
- Increased time or number of pulses results in smoother trace and increased accuracy
- Use high enough value to produce smooth trace

© Pearson Technologies 10- Testing and Inspection 156

156

Backscatter Coefficients

- Correct value makes attenuation rate values accurate

© Pearson Technologies 10- Testing and Inspection 157

157

Multimode Backscatter Coefficients

Fiber	850 nm	1300 nm
OFS 550/300	-68.4 dB	-75.8 dB
OFS LaserWave G+	-68.4 dB	-75.8 dB
Corning InfiniCor® 62.5	-68 dB	-76 dB
Corning InfiniCor® 50	-68 dB	-76 dB

© Pearson Technologies 10- Testing and Inspection 158

158

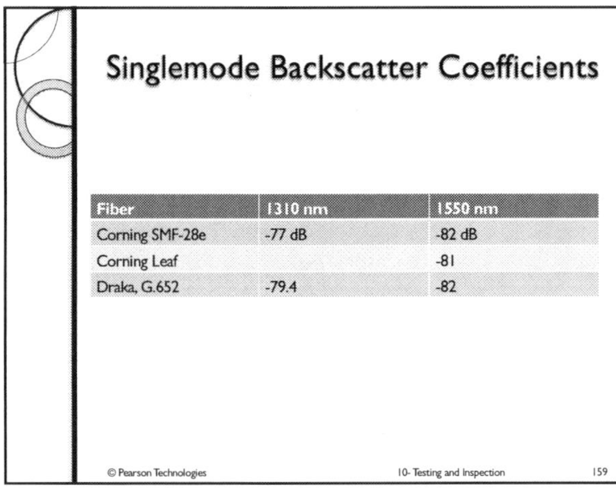

Singlemode Backscatter Coefficients

Fiber	1310 nm	1550 nm
Corning SMF-28e	-77 dB	-82 dB
Corning Leaf		-81
Draka, G.652	-79.4	-82

© Pearson Technologies 10- Testing and Inspection 159

159

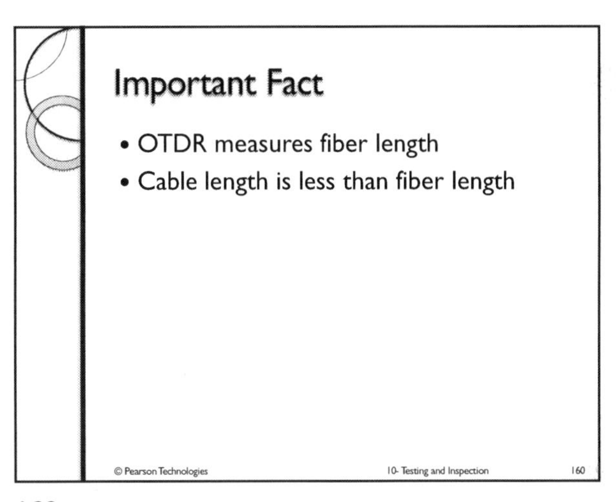

160

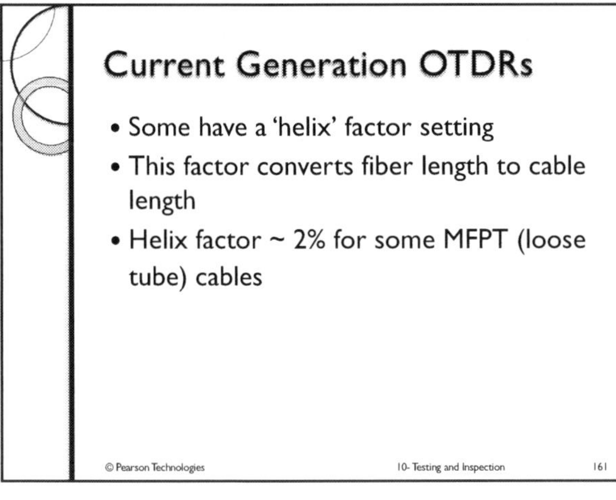

161

Example: Distance Inaccuracies

Distance, m		100	500	1000	5000	10000
fiber	buffer tube					
Excess length	Excess length	Distance error, m				
0.01%	1.91%	1.92	9.60	19.19	95.97	191.94
0.02%	1.91%	1.93	9.65	19.29	96.47	192.94
0.03%	1.91%	1.94	9.70	19.39	96.97	193.94
0.04%	1.91%	1.95	9.75	19.49	97.47	194.94
0.05%	1.91%	1.96	9.80	19.59	97.97	195.94
0.06%	1.91%	1.97	9.85	19.69	98.47	196.94
0.07%	1.91%	1.98	9.90	19.79	98.97	197.94
0.08%	1.91%	1.99	9.95	19.89	99.47	198.94
0.09%	1.91%	2.00	10.00	19.99	99.97	199.94
0.10%	1.91%	2.01	10.05	20.09	100.47	200.94

659 FT

© Pearson Technologies 10- Testing and Inspection 162

162

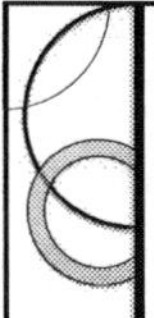

Always Use Launch Cable

- To measure the loss of the near end connector
  - Launch cable must be longer than dead zone width to measure loss of near end connector
- To protect the OTDR port
- For multimode cables, launch cable can be longer than longest cable to measure to avoid ghost reflections

163

Trace With Launch Cable

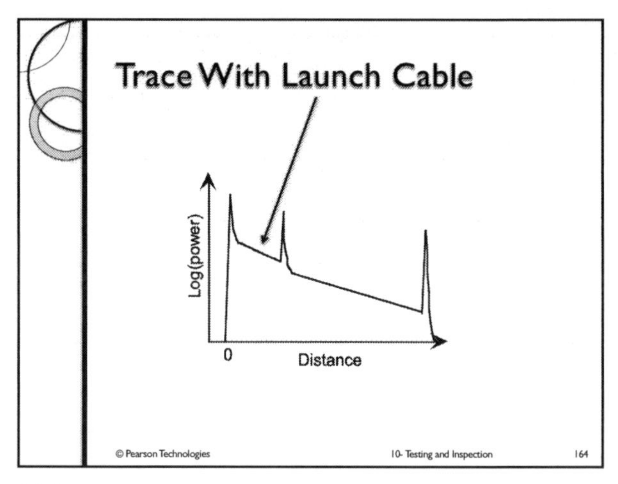

164

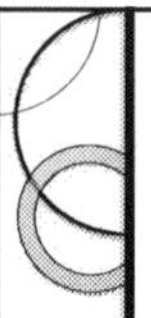

Launch Cable Length

- Some OTDRs allow entry of this length
- With this length, all measurements are adjusted to ignore this length
- With this feature, testing with a different launch cable does not require adjusting original measurements for new launch cable length

165

Part 2: OTDR Testing

- General Information
- Purpose
- Types
- Principles of Operation
- Information Produced
- Three Basic Traces
- Using the OTDR
  - Setting Up
  - Making measurements
- Manual Measurements

166

Read next slide carefully.

KEY UNDERSTANDING

167

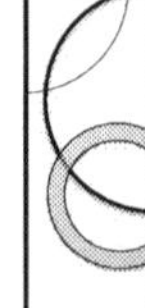

2 Reasons for Manual Measurements

- Verify accuracy of software
  - Once done, trust software
  - Manual measurements unnecessary
- However, OTDR cannot accurately interpret loss within dead zone
  - OTDR does not have internal map that indicates the configuration within any dead zone
  - Software has a single value for dead zone, independent of number of events within
  - Manual interpretation necessary for accurate interpretation of loss within dead zone
- Proper interpretation of trace requires map
  - Did you hear an echo? (Third mention?)

168

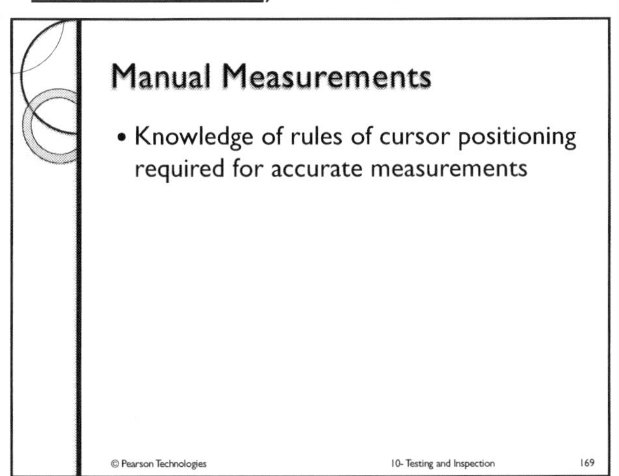

169

Cursor Placement Rules

- Ensure accurate measurements
- Accurate measurements lead to proper interpretations

© Pearson Technologies 10- Testing and Inspection 170

170

Measurements To Make

- Length
  - First segment
  - Subsequent segments
  - Distance to faults
- Connector and splice loss
  - Estimated method
  - Accurate method
- Attenuation rate

© Pearson Technologies 10- Testing and Inspection 171

171

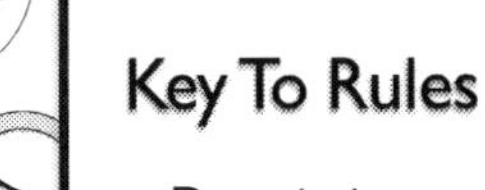

Key To Rules

- Descriptions are in black
- Rules are in red
- Important Rule
  - Cursor is never placed in a dead zone (a peak or drop off)

© Pearson Technologies 10- Testing and Inspection 172

172

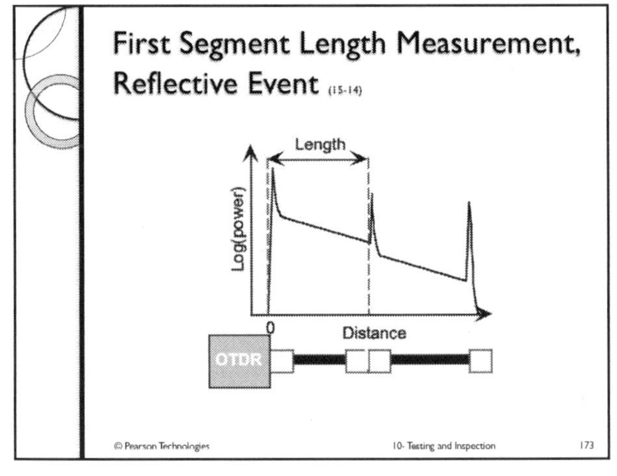

173

Rule 1: First Segment Length

- The end of the segment is defined by the lowest point of the backscatter trace before the peak
- One cursor required
- Place cursor is at lowest point of straight-line trace before a peak

© Pearson Technologies 10- Testing and Inspection 174

174

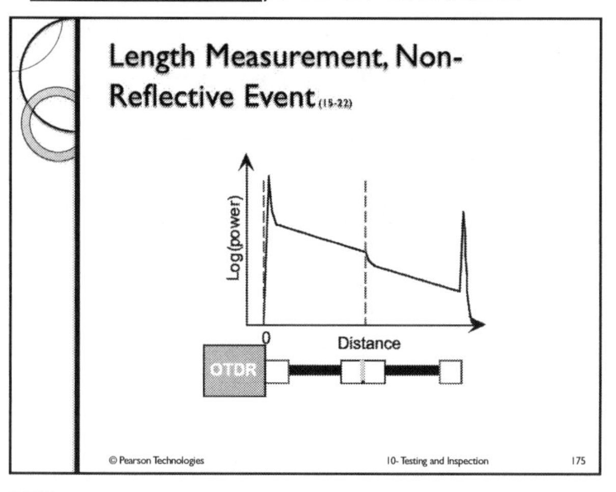

175

Rule 2: First Segment Length

- The end of the segment is defined by the lowest point of the backscatter trace before the change in slope, or drop off, that marks the end of the first segment
- Place cursor is at lowest point of straight-line trace before the change in slope, or drop off, that marks the end of the first segment

© Pearson Technologies 10- Testing and Inspection 176

176

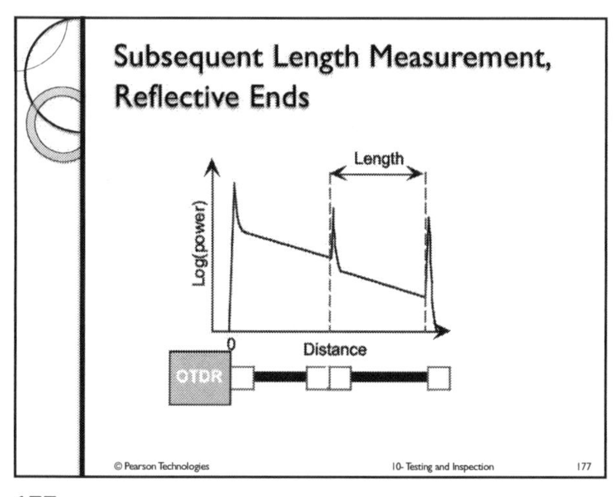

177

Rule 3: Subsequent Segment Length

- Two cursors required
- Place one cursor at the lowest point on the straight-line trace before the peak that marks the end of the previous segment
- Place the second cursor at the lowest point of a straight-line trace of the segment before either the peak that marks the end of that segment or the non-reflective drop that marks the end of the segment

© Pearson Technologies 10- Testing and Inspection 178

178

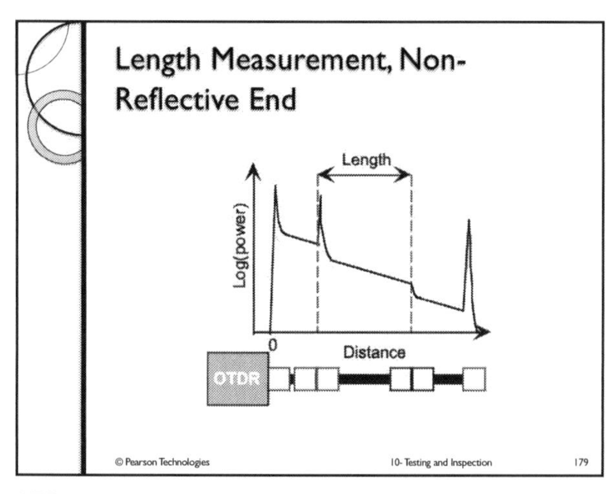

179

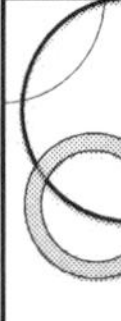

Rule 4: Subsequent Segment Length, Non-Reflective Ends

- Two cursors required
- Place the first cursor at the lowest point on the straight-line trace before the peak or drop that marks the end of the previous segment
- Place the second cursor at the lowest point of a straight-line trace of the segment being measured before the peak or drop that marks the end of that segment

© Pearson Technologies 10- Testing and Inspection 180

180

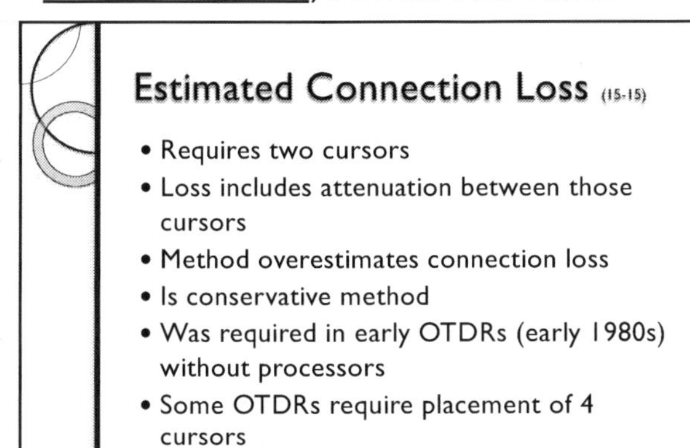

181

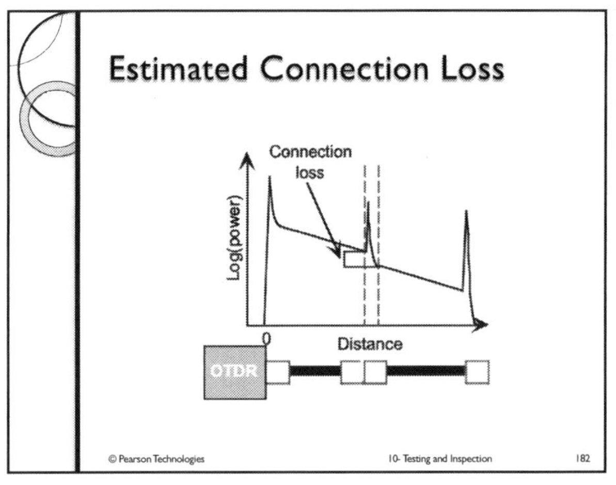

182

Rule 5: Estimated Connection Loss

- Place cursers on both sides of the peak or drop that indicates the connection
- The cursers are in the straight-line traces on both sides of the connection
- The cursers are as close as possible to the dead zone without being in the dead zone (peak or drop)

© Pearson Technologies 10- Testing and Inspection 183

183

Incorrect, High Loss #1

Connection loss

Log(power)

0

Distance

OTDR

© Pearson Technologies 10- Testing and Inspection 184

184

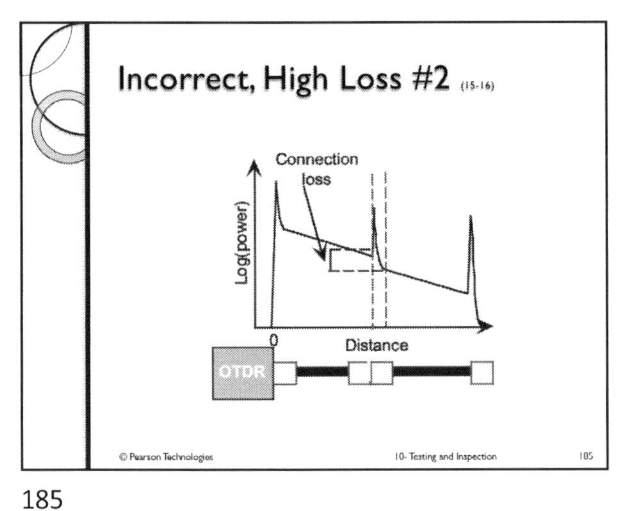

185

Incorrect, Low Loss

Connection loss

Log(power)

0

Distance

OTDR

© Pearson Technologies 10- Testing and Inspection 186

186

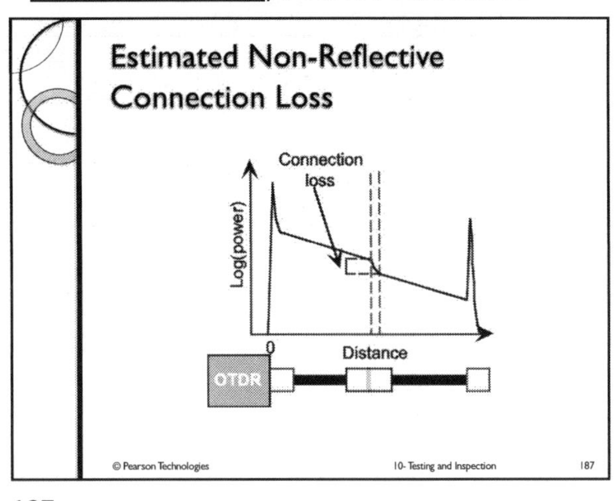

187

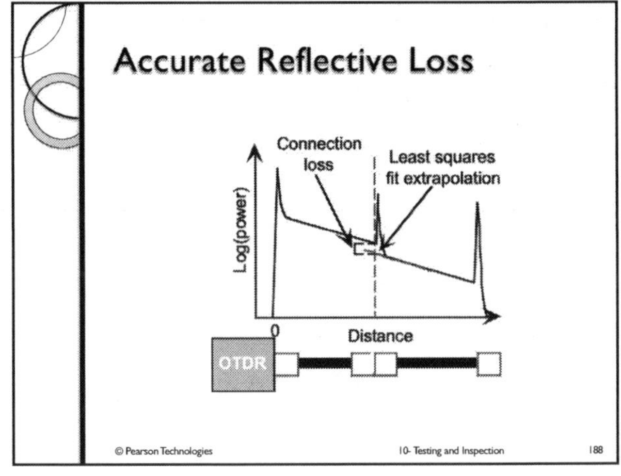

188

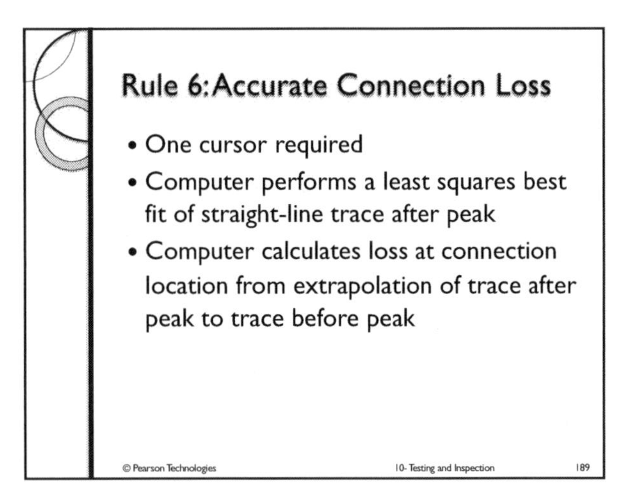

189

Accurate Non-Reflective Loss (15-17)

Connection loss

Least squares fit extrapolation

Log(power)

0

Distance

OTDR

© Pearson Technologies 10- Testing and Inspection 190

190

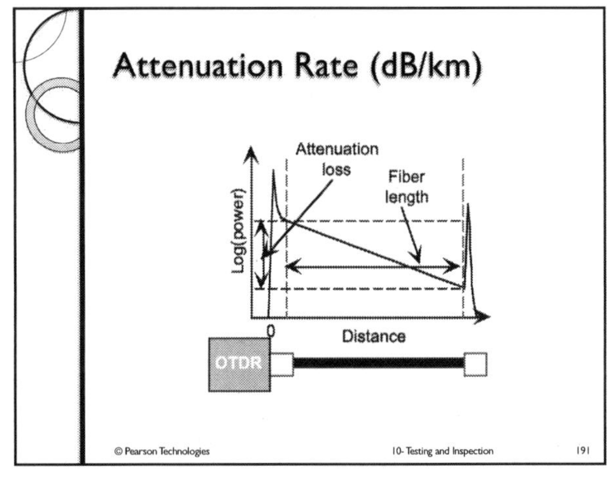

191

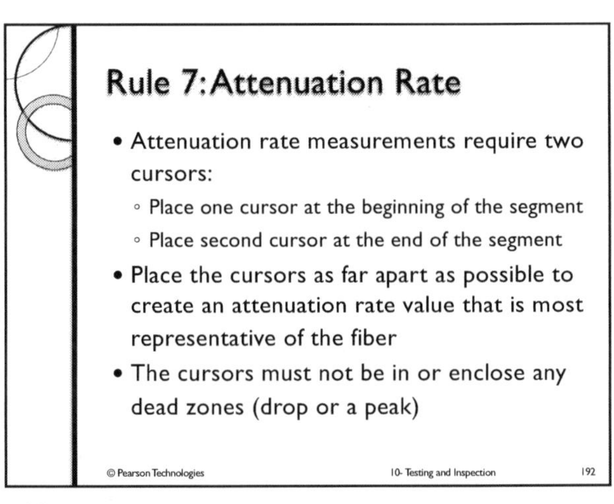

192

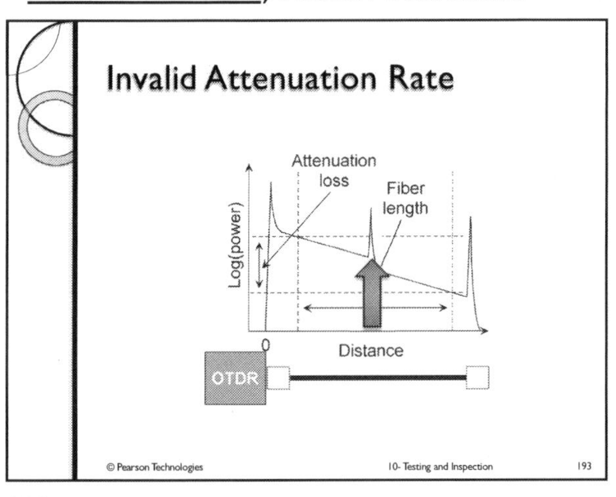

193

Acceptable Placement

- To avoid placing cursors in dead zones, move cursors closer to each other
- Attenuation rate value will change
- Interpretation will not change!

© Pearson Technologies 10- Testing and Inspection 194

194

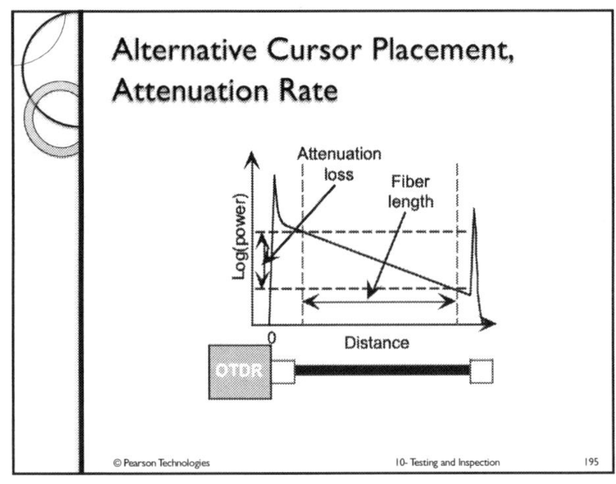

195

Important Reminder

- Do not place cursors in dead zones
- There are no valid measurements to be made with cursors in dead zones
- Dead zones are created by OTDR, not by link components

© Pearson Technologies 10- Testing and Inspection 196

196

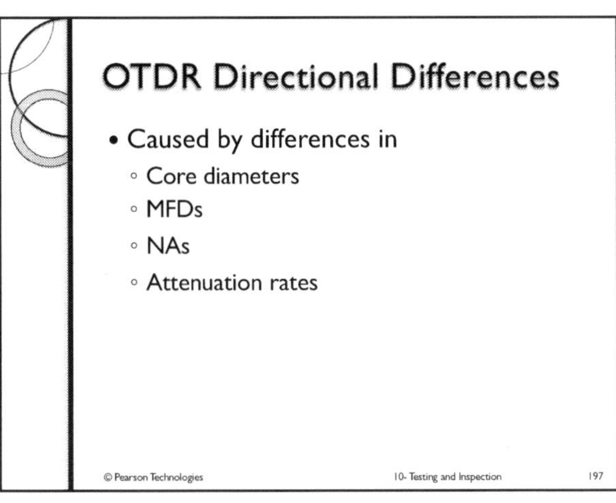

197

OTDR Test Summary

- Number of wavelengths
  - 2 necessary to verify stress free installation
- Number of directions
  - 2 necessary for true splice loss
  - 2 necessary to identify ghosts as ghosts

© Pearson Technologies 10- Testing and Inspection 198

198

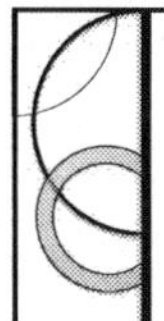

OTDR Limitations

- Although the OTDR provides highly useful information, there are significant limitations
  - Features in dead zone are invisible
  - Limited distance resolution, created by the dead zone length, may limit usefulness in premises applications

© Pearson Technologies 10- Testing and Inspection 199

199

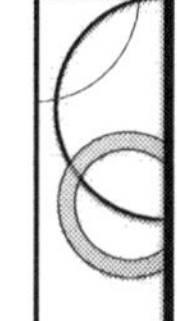

OTDR Limitations from Field Work

- Inaccurate and non-reproducible attenuation rates from short, singlemode segments
  - 'Short' can extend to ~ 2000m
- Excessive attenuation rates from short segments of bend insensitive multimode from some manufacturers

© Pearson Technologies 10- Testing and Inspection 200

200

Hands-On OTDR Evaluation

- Do not proceed until you hear the magic word: Proceed
- Attach connector to launch cable
- Set wavelength, time, pulse width and maximum length
- Start test
- Observe trace
- Does each segment have uniform loss?
- Record all results
- Place cursors for attenuation rate measurement
- Place cursors for connector loss measurement
- Place cursors for splice loss measurement
- Place cursors for segment length measurement

© Pearson Technologies 10- Testing and Inspection 201

201

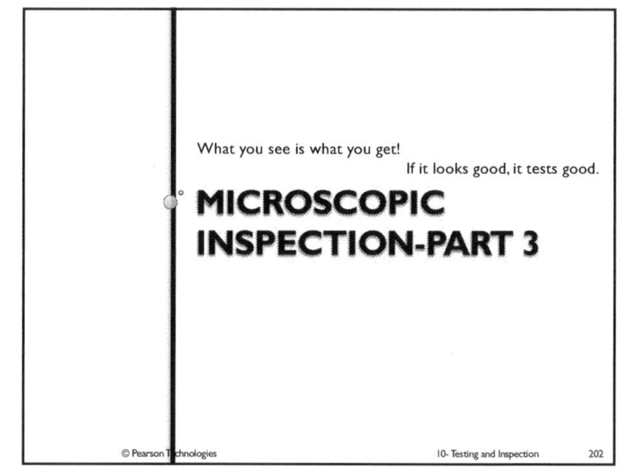

What you see is what you get!
If it looks good, it tests good.

MICROSCOPIC INSPECTION-PART 3

© Pearson Technologies 10- Testing and Inspection 202

202

Part 3: Microscopic Inspection

- Learn how to inspect connectors
- Learn how to rate or interpret microscopic appearances
- Learn how to identify corrective actions

© Pearson Technologies 10- Testing and Inspection 203

203

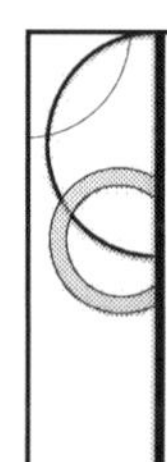

Applicability

- This part applies to
  - All connectors that require field polishing
  - Connectors that have been in use

© Pearson Technologies 10- Testing and Inspection 204

204

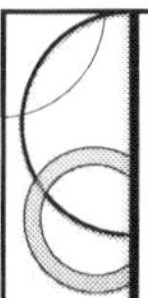

Reason For Inspection

- A 'good' appearance correlates with low loss, most of the time
- Cleave and crimp connectors may not exhibit this correlation, because loss is determined by quality of cleaved fiber ends inside the connectors

© Pearson Technologies 10- Testing and Inspection 205

205

Equipment Required

- A 400-magnification connector inspection microscope with an IR filter
  - Some professionals recommend 100x and 200x microscopes
  - Recommended: www.Westoverscientific.com
- Lens grade, lint free, tissues (Kim wipes or equivalent)
- Electro-Wash® Px (recommended)
  - Alternative: connector cleaner without isopropyl alcohol (opinion)

© Pearson Technologies 10- Testing and Inspection 206

206

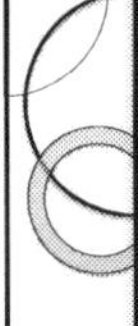

Microscope inspection

© Pearson Technologies 10- Testing and Inspection 207

207

Inspection Probe

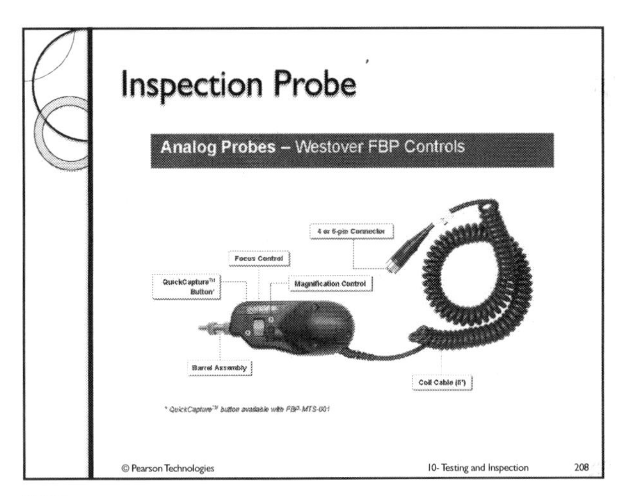

© Pearson Technologies 10- Testing and Inspection 208

208

Procedure

- Remove cap
- Clean connector
- Install connector
- Focus
- View and rate connector
- Repeat viewing and rating with back light

© Pearson Technologies 10- Testing and Inspection 209

209

Back Light

- Is a white light launched into opposite end of fiber
- Back light can reveal features
  - Example: fiber broken below surface but in ferrule
- Back light can conceal features
- Whenever possible, inspect both ways

© Pearson Technologies 10- Testing and Inspection 210

210

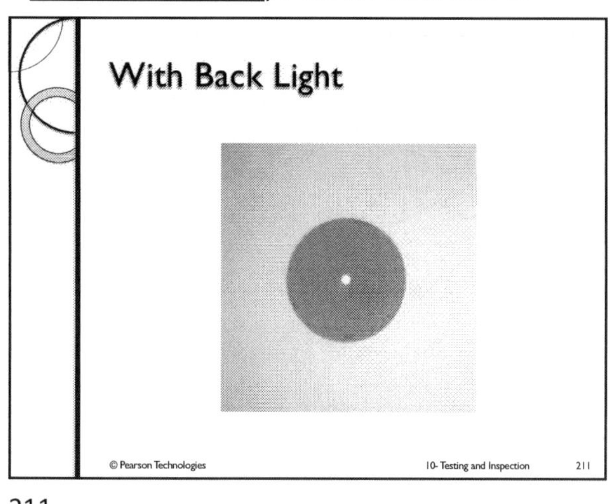

211

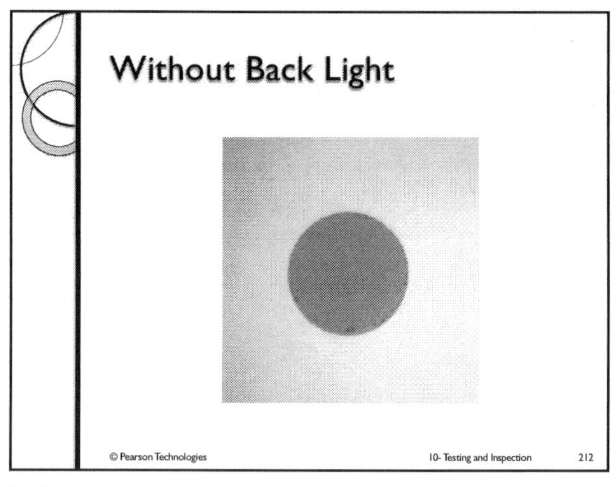

212

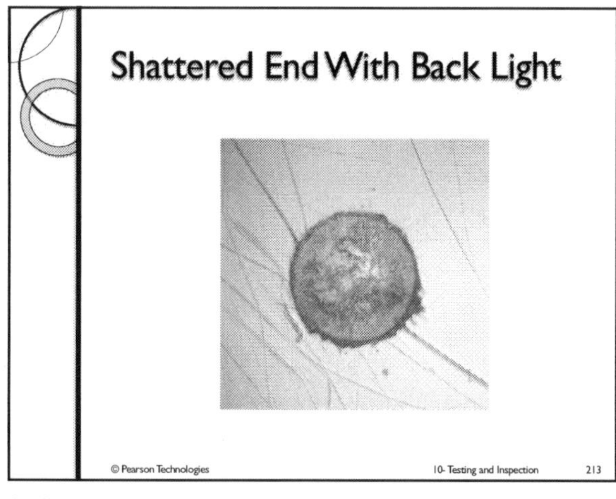

213

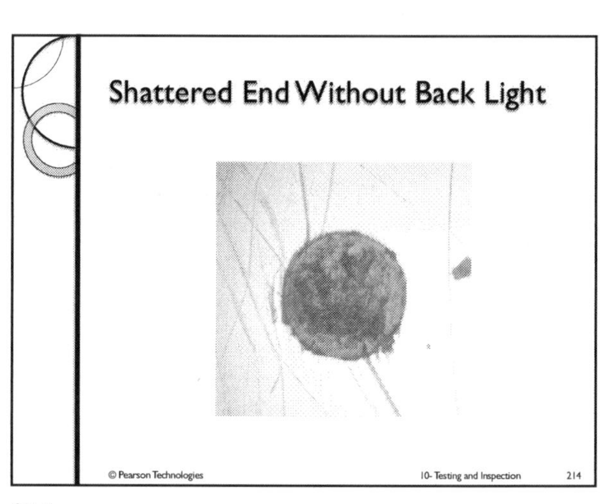

214

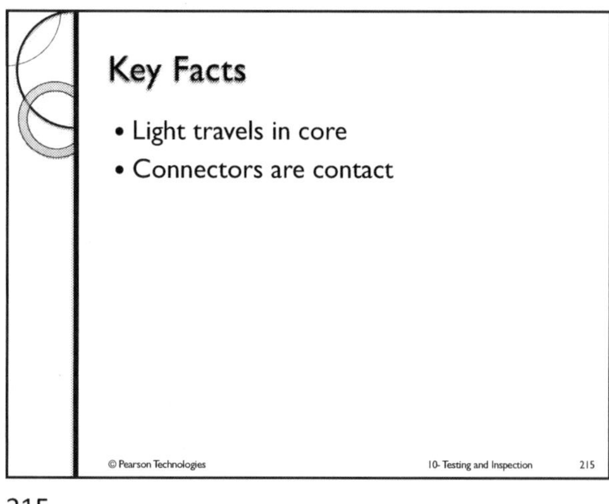

215

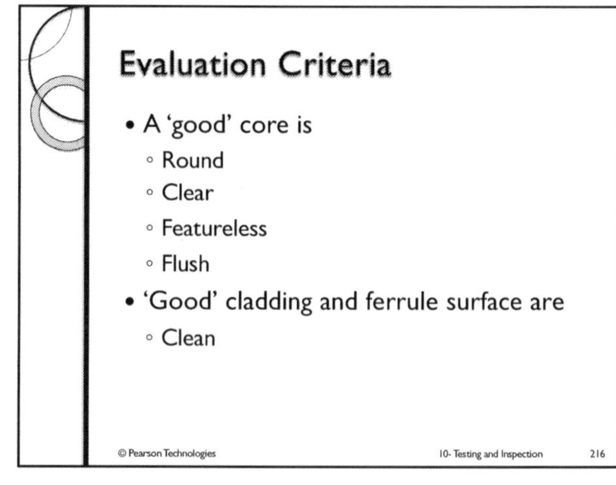

216

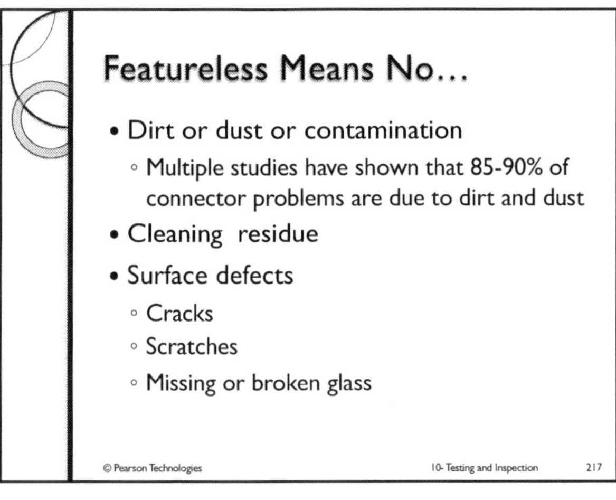

217

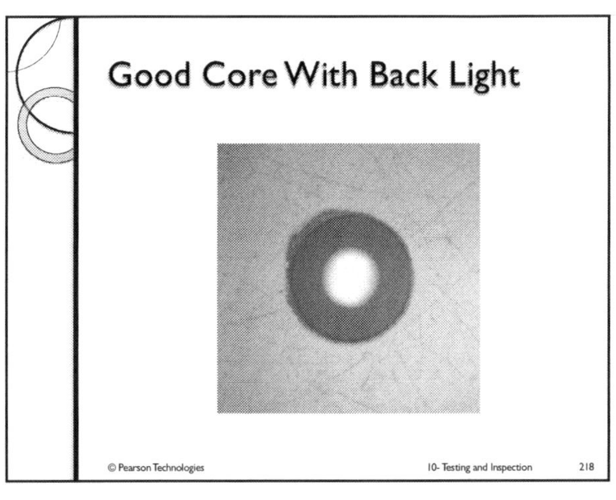

218

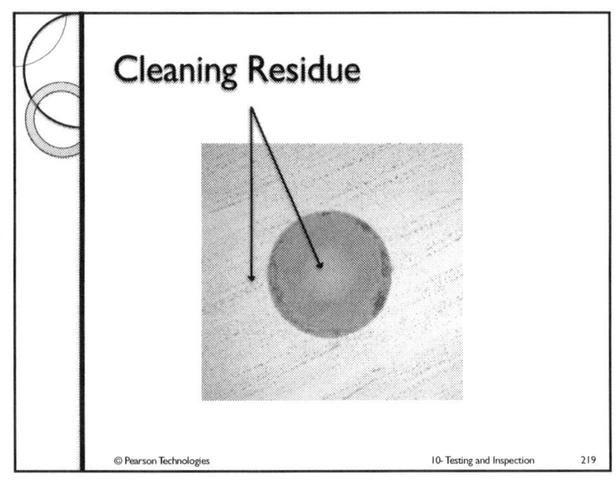

219

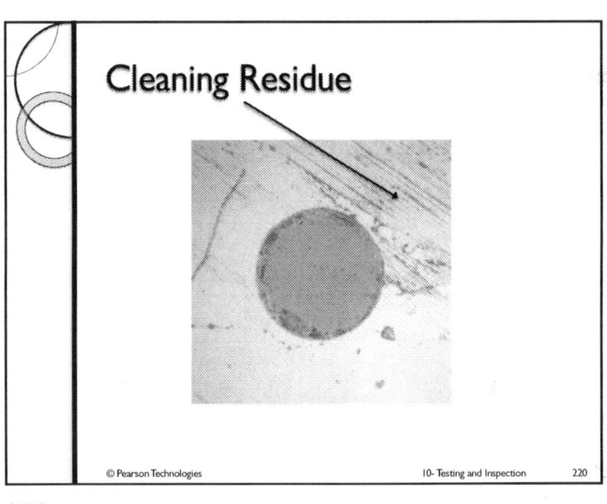

220

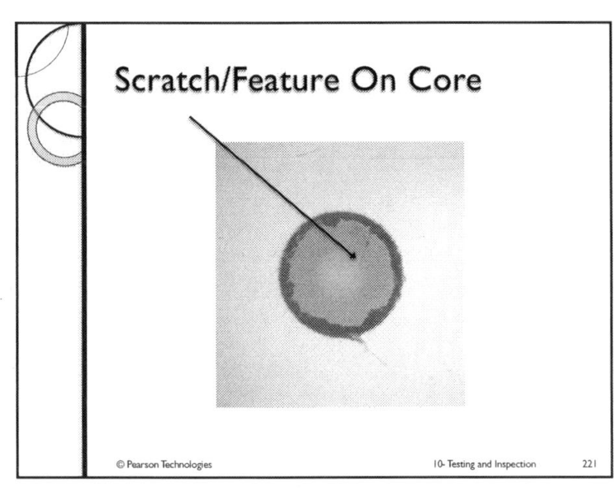

221

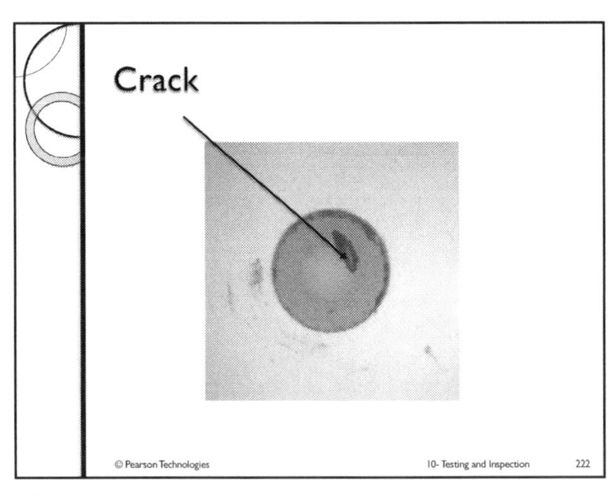

222

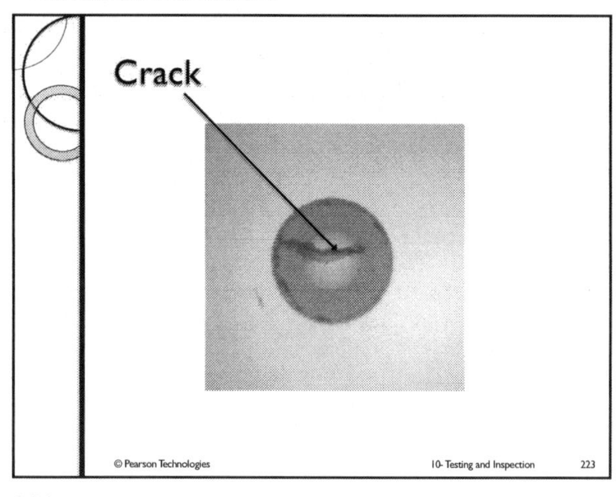

223

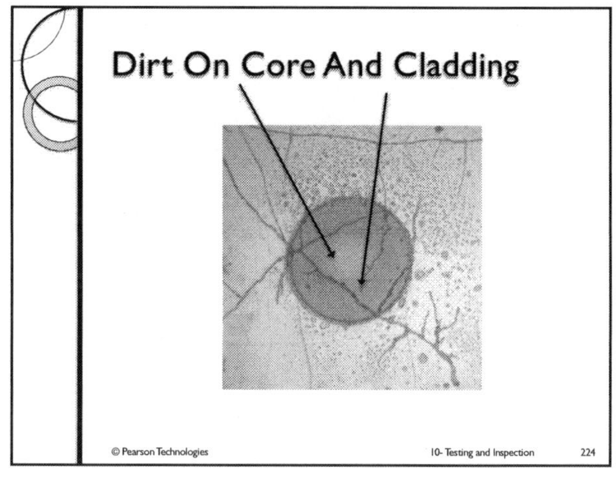

224

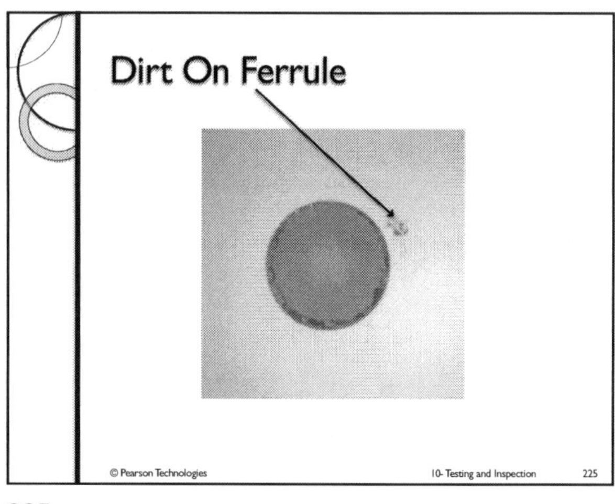

225

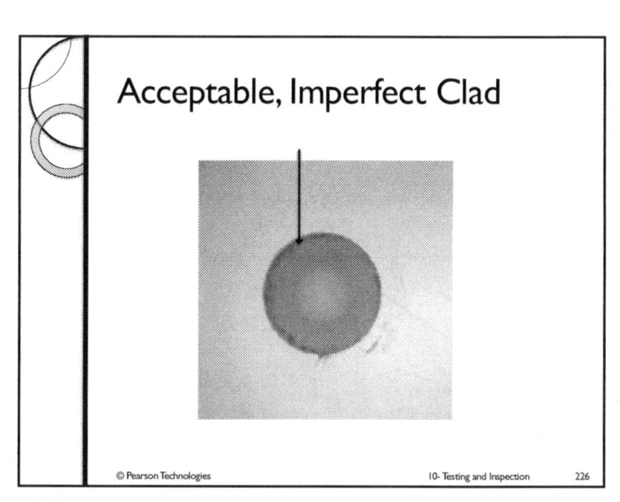

226

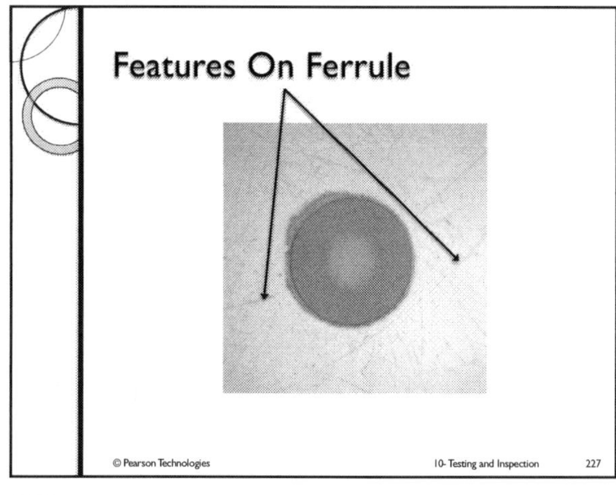

227

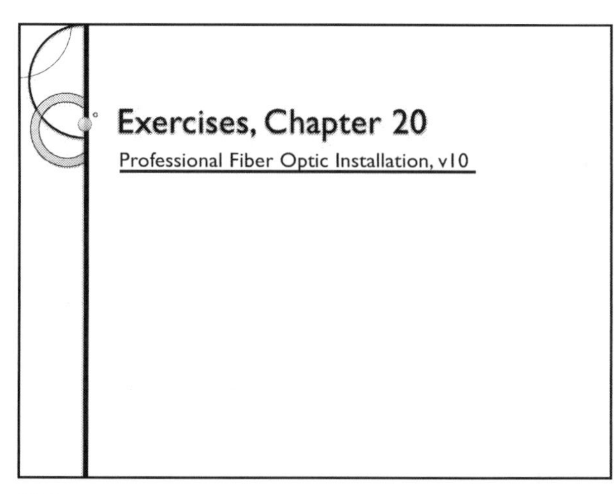

228

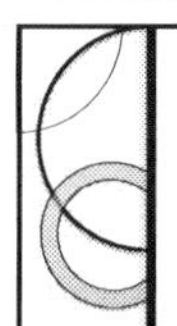

Hands-On Microscopic Inspection

- Do not proceed until you hear the magic word: Proceed
- Clean connector
- Place connector in microscope or video microscope
- Focus
- Inspect and rate connector
- Repeat with back light
- Record evaluation

© Pearson Technologies 10- Testing and Inspection 229

229

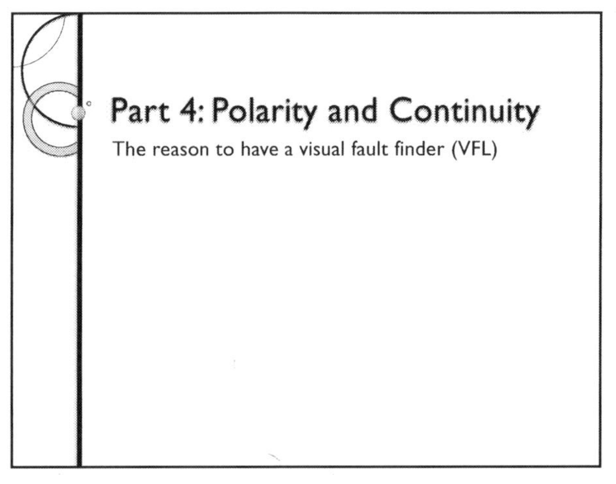

Part 4: Polarity and Continuity

The reason to have a visual fault finder (VFL)

230

Visual Feature/Fault Locator

- The visual feature/fault locator is a high-power red laser with visible wavelength of ~670nm
- The VFL is used to
  - Verify continuity
  - Verify polarity
  - Reveal locations of power loss
    - In connectors
    - In splices
    - In fibers and buffer tubes

© Pearson Technologies 10- Testing and Inspection 231

231

VFL

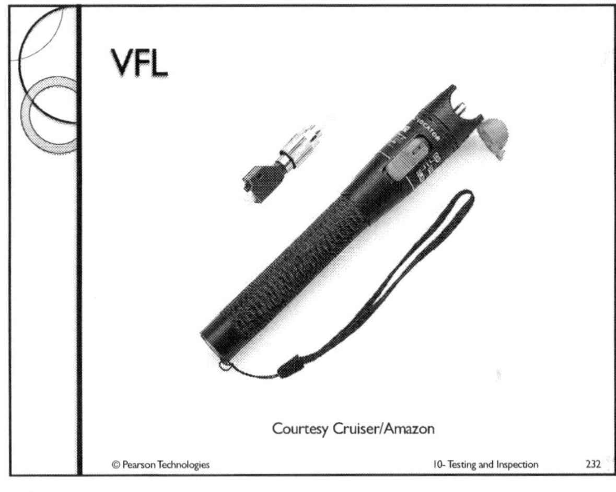

Courtesy Cruiser/Amazon

© Pearson Technologies 10- Testing and Inspection 232

232

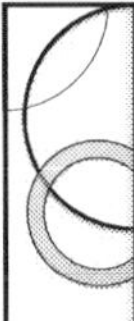

Power Loss Revealed by

- Glow at location indicating power escaping (leaking) from core
  - In tight tube from broken fiber
  - In backshell of connector
  - In fiber within splice cover
  - Within enclosure due to bend radius violation

© Pearson Technologies 10- Testing and Inspection 233

233

VFL Is Over Test

- Low to medium glow expected
  - VFL launches light in cladding
  - Cladding light escapes
  - Low to medium glow can be low loss
  - Replace splice or connector only after OTDR verification of high loss
- High glow (like a Christmas tree light) indicates high loss
  - Replace splice or connector

© Pearson Technologies 10- Testing and Inspection 234

234

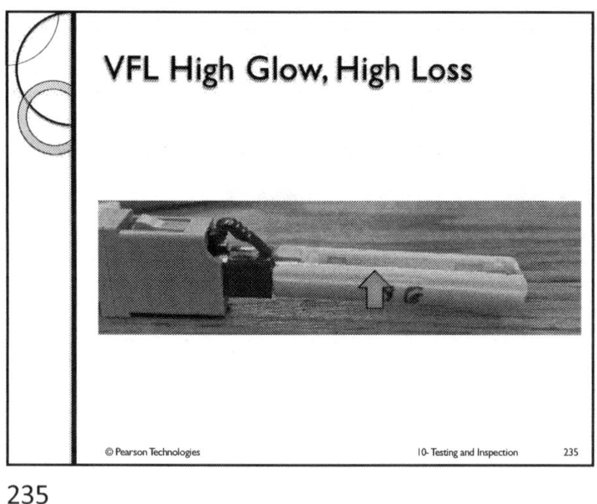

235

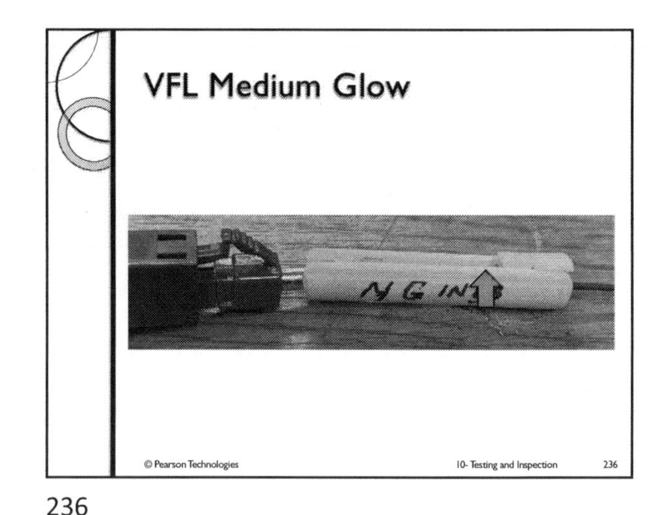

236

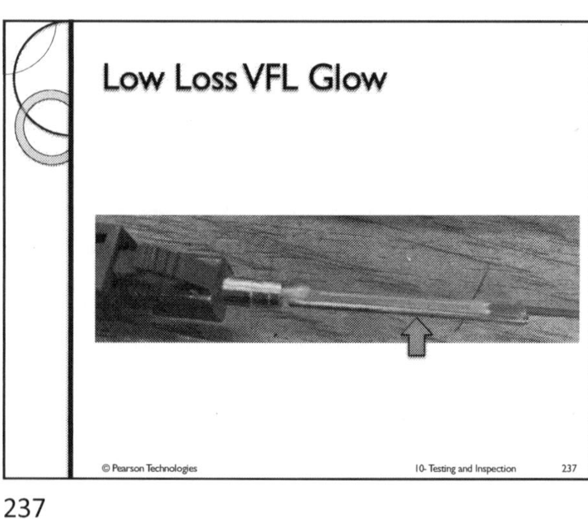

237

Interpret Cautiously

- VFL launches light into core and cladding
- Cladding light can travel significant distance, indicating loss where there is none
- VFL interpretation requires recognizing the over test condition
  - All that glows is not 'bad'
  - Interpretation of cladding glow requires experience

© Pearson Technologies 10- Testing and Inspection 238

238

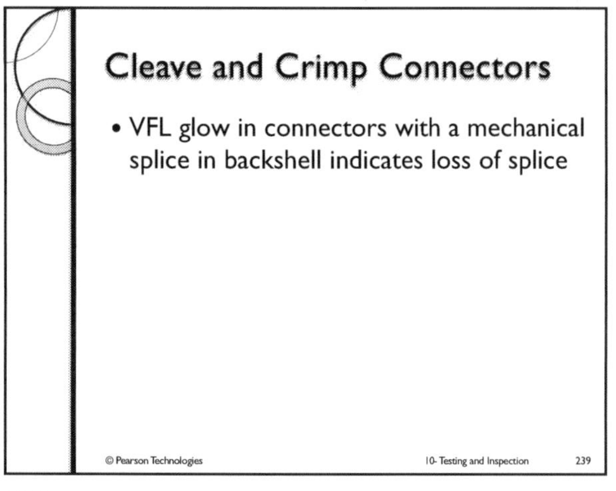

239

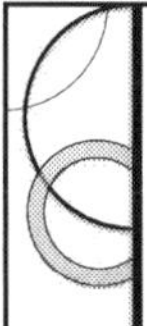

Hands-On VFL Evaluation

- Do not proceed until you hear the magic word: Proceed
- Place connector in VFL
- Turn on VFL
- Inspect opposite end
- Continuous?
- Reverse VFL end and inspect far end
- Does connector glow?
- Record evaluation

© Pearson Technologies 10- Testing and Inspection 240

240

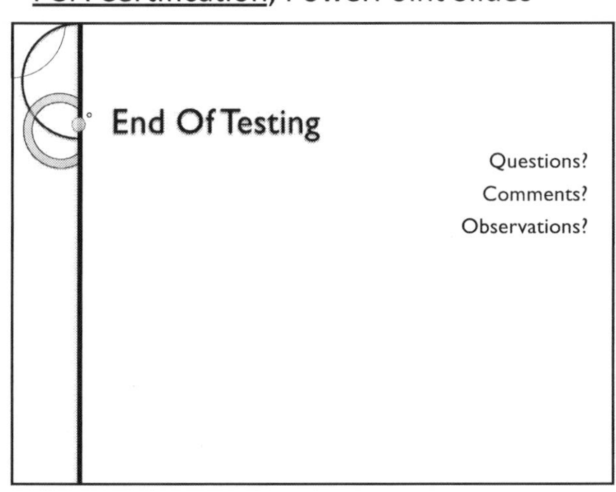
End Of Testing
Questions?
Comments?
Observations?

241

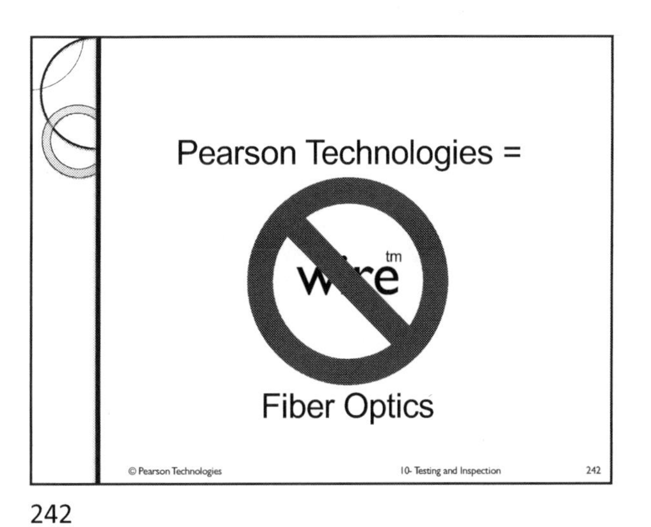
Pearson Technologies =
wire
tm
Fiber Optics
© Pearson Technologies
10- Testing and Inspection
242

242

CFOT Certification Examination Preparation Questions

INSTRUCTIONS FOR USE

After each section of the PowerPoint presentation, answer the questions indicated. For additional practice, answer the review questions in the FOA practice quizzes. Find these quizzes at the web addresses at the end of this appendix.

Forms of these 141 questions will be on the FOA Certified Fiber Optic Technician 101 question certification examination. Preparation with these questions and with the practice quizzes will enable you to pass the certification examination.

AFTER SECTION 1

From Jargon Quiz Page 1-15

6. Loss of a fiber or any fiber in a cable is measured in __________.
   A. dB
   B. dBm
   C. milliwatts

7. 10 dB corresponds to a factor of __________in power.
   A. 2
   B. 10
   C. 20
   D. 100

From Datalink Quiz Page 2-9

7. Fiber amplifiers and DWDM work in the __________wavelength range.
   A. 650-850
   B. 850-1300
   C. 1300-1550
   D. 1480-1650

11. The wavelength of light used for most fiber optic systems is in the _____________region and ____to the human eye.
   A. ultraviolet, invisible
   B. solar, visible
   C. infrared, invisible

AFTER SECTION 2

FROM DATA LINK QUIZ page 2-9

_2. LEDs have higher output power and bandwidth than lasers. [T/F]

Multiple Choice

Identify the choice that best completes the statement or answers the question.

_3. Multimode fiber systems operation at speeds of 1 Gb/s or more use ________sources.

A. LED
B. VCSEL
C. F-P laser
D. DFB laser

_4. The ____________of a laser makes the effective bandwidth of multimode fiber higher than with LEDs.

A. Restricted modal launch
B. Higher power
C. Lower power
D. Bandwidth

5. Short wavelength 850 nm links can use __________detectors in the receiver.

A. Silicon
B. Germanium
C. InGaAs

6. Long wavelength singlemode links at wavelengths in the range of 1300-1650 nm links must use ______detectors in the receiver for the best sensitivity performance.

A. Silicon
B. Germanium
C. InGaAs

7. Fiber amplifiers and DWDM work in the __________wavelength range.

E. 650-850
F. 850-1300
G. 1300-1550
H. 1480-1650

FROM DESIGN 2 QUIZ PAGE 7-18

Multiple Response

Identify one or more choices that best complete the statement or answer the question.

8 Singlemode transceivers use ______sources for their higher coupled power and bandwidth.

A. LED
B. VCSEL
C. F-P lasers
D. DFB lasers

9 Multimode transceivers use ______sources depending on their requirements for coupled power and bandwidth.

a. LED
b. VCSEL
c. F-P lasers
d. DFB lasers

CFOT Certification Examination Preparation Questions

AFTER SECTION 3

From Jargon Quiz Page 1-15

1. Optical fibers can transmit either analog or digital signals. [T/F]

From Datalink Quiz Page 2-9

1. Fiber optic links generally use two fibers for full duplex (bidirectional) links. [T/F]

From Basics Page 3-9

1. Most outside plant installations are singlemode fiber. [T/F]

2. Splicing is very rare in premises networks. [T/F]

3. Fiber is used in long distance phone networks because it is much cheaper than copper wire. [T/F]

4. Dangerous light from fiber optic cables is bright and easily visible. [T/F]

5. Besides causing attenuation, dirt particles can cause scratches on the polished fiber ends. [T/F]

6. Outside plant cabling can be installed by ___________________.
   A. Pulling in underground in conduit
   B. Direct burial
   C. Aerial suspension
   D. All of the above

7. Underground cable generally includes a gel, powder or tape for protection from
   A. Pulling friction
   B. Lightning strikes
   C. Moisture
   D. Fiber abrasion

8. Armored cable is used in outside plant installations to ___________________.
   A. Prevent rodent damage
   B. Protect from dig-up damage
   C. Increase pulling tension
   D. Conduct lightning strikes

10. Premise cables in LAN backbones often contain _______________.
    A. Only multimode fiber
    B. Only singlemode fiber
    C. Both multimode and singlemode fiber
    D. Plastic optical fiber

13. Information on the safety of chemicals used in fiber optics are ________________.
    A. Available from National Institutes of Health
    B. In MSDS sheets supplied by manufacturers
    C. Required to be in every installer's tool kit
    D. Rarely useful

14. Always keep _______________on connectors when not connected to equipment or being tested.

A. Mating adapters
B. Strain relief boots
C. Sticky tape
D. Dust caps

From BASIC QUIZ, page 3-9

___ 12. The protective gear every VDV installer must always wear is ________

A. Eye protection
B. Plastic apron
C. Gloves
D. Shoe covers

From Networks Quiz Page 2-9

1. The biggest advantage of optical fiber is the fact it is the most cost effective means of transporting information. [T/F]

2. Telephone networks have been converted to fiber, including long distance and metropolitan networks, but fiber to the home (FTTH) is not yet feasible. [T/F]

Identify the choice that best completes the statement or answers the question.

3. In an industrial environment, fiber is most often used to
   A. Prevent electromagnetic interference
   B. Provide ultra high-speed connections to machines
   C. Withstand high temperatures
   D. Tolerate physical abuse

4. Which of the following are not necessary in a centralized fiber optic cabling architecture?
   A. Repeaters or hubs
   B. Telecom closets
   C. Wall outlets
   D. NIC cards

5. Copper networks can be converted to fiber optics easily using:
   A. Fiber hubs
   B. Media converters
   C. Patch panels
   D. Rewiring

6. The bandwidth and distance capability of optical fiber means that ____________. (choose all that apply)
   A. Fewer cables are needed
   B. Fewer repeaters are needed
   C. Less power is consumed by the network
   D. Less maintenance is required

7. Which of the following typically use fiber optic backbones? (choose all that apply)
   E. Telephones
   F. CATV
   G. Internet
   H. Cell Phones

CFOT Certification Examination Preparation Questions

From Installation Quiz Page 10-41

1. Industrial applications often use fiber optics for its noise immunity rather than distance and bandwidth advantages.

5. Outside plant cabling can be installed by ____________.
   A. Pulling in underground in conduit
   B. Direct burial
   C. Aerial suspension
   D. All of the above

6. The bandwidth and distance capability of optical fiber means that _________. (choose all that apply)
   A. Fewer cables are needed
   B. Fewer repeaters are needed
   C. Less power is consumed by the network
   D. Less maintenance is required

7. Which of the following typically use fiber optic backbones? (choose all that apply)
   A. Telephones
   B. CATV
   C. Internet
   D. Cell Phones

11. The industry standard that covers structured cabling, both fiber and copper, is __________.
    A. TIA-568
    B. TIA-526-14
    C. IEEE 802.3
    D. NECA-301

12 Structured cabling installed to TIA-568 standards uses a ___________cabling architecture.
    A. Bus
    B. Ring
    C. Star
    D. Tree

AFTER SECTION 4

From JARGON Quiz, PAGE 1-15

2. Singlemode fiber has a smaller core than multimode fiber. [T/F]

Multiple Choice
Identify the choice that best completes the statement or answers the question.

3. In an optical fiber, the light is transmitted through the _______________.
   A. Core
   B. Cladding
   C. Buffer
   D. Jacket

4. The diameter of an optical fiber is traditionally measured in __________.
   A. Meters

B. Millimeters
C. Microns (micrometers)
D. Nanometers

5. Rays of light transmitted in multimode fiber are called ____________.
   A. Reflections
   B. Refractions
   C. Waves
   D. Modes

8. A fiber stripper removes the ______________of the fiber.
   A. Core
   B. Cladding
   C. Buffer coating

9. The ________ protects the fiber from harm.
   A. Primary buffer coating
   B. Aramid fiber strength members
   C. Jacket
   D. All of the above

From INSTALLATION QUIZ, page 9-16

__ 12. Structured cabling installed to TIA-568 standards uses a _________cabling architecture.
   A. Bus
   B. Ring
   C. Star
   D. Tree

From Fiber Quiz Page 4-14

1. Singlemode fiber has a ___________light-carrying core than multimode fiber.
   A. Smaller
   B. Larger
   C. Same size

2. What is the core size of singlemode fiber?
   A. 5 mm
   B. 9 microns
   C. 50 microns
   D. 63.5 microns

3. Singlemode fiber has _____________bandwidth than multimode fiber.
   A. More
   B. Less
   C. The same

4. What wavelengths are appropriate for use with multimode fiber?
   A. 650 & 850 nm
   B. 850 & 1300 nm
   C. 850 & 1310 nm
   D. 1310 & 1550 nm

_5. The diameter of the core in OM2 and OM3 multimode fiber is how large?

A. 50 microns
B. 62.5 microns
C. 62.5 mm
D. 9 mm

6. Which of the following fiber specifications is most important to the user and is an important factor in testing?
A. Attenuation
B. Bandwidth
C. Numerical aperture
D. Core-cladding concentricity

7. The largest contributor to fiber attenuation is ________________.
A. Absorption
B. Scattering
C. Bending losses
D. Micro-bends

8. Which fiber typically has the largest core?
A. POF
B. Multimode Step Index
C. Multimode Graded Index
D. Singlemode

9. The loss of a multimode graded index fiber is greatest at ______________.
A. 850 nm
B. 1300 nm
C. 1310 nm
D. 1550 nm

10. Which type of dispersion affects singlemode fiber as well as multimode fiber?
A. Modal
B. Differential
C. Chromatic
D. Polarization mode

From DESIGN 2 Quiz, page 7-18

7. The first requirement that must be considered for a new fiber optic project is ___________.
A. The customer's communications system requirements
B. Where the cable plant will be run
C. Whether it will be multimode or singlemode fiber
D. The customer's budget

From DESIGN SELF STUDY QUIZ, page 7-18

8. When calculating the contribution of the fiber loss to the loss budget, one must consider the ______________________________.

A. Size of the fiber
B. Type of cable
C. Termination of the fiber
D. Wavelength of the light in the fiber

AFTER SECTION 5

From JARGON QUIZ, page 1-15

9. The _________ protects the fiber from harm.
   A. Primary buffer coating
   B. Aramid fiber strength members
   C. Jacket
   D. All of the above

From CABLE QUIZ, page 5-28

Matching

Identify the cable types

A

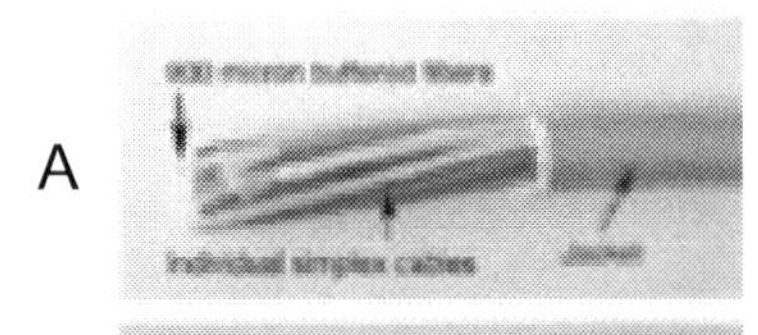

B

C

D

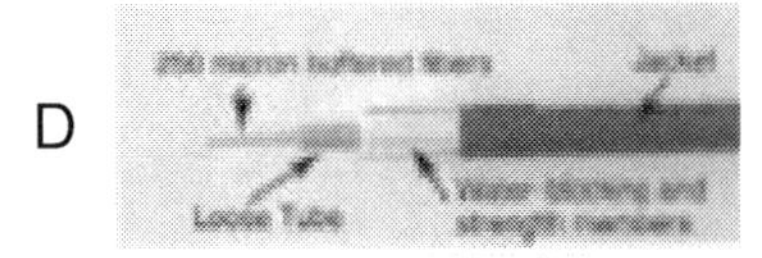

E

__ 1. Zipcord

__ 2. Distribution

__ 3. Loose tube

__ 4. Ribbon

__ 5. Breakout

CFOT Certification Examination Preparation Questions

True/False
Indicate whether the statement is true or false.

6. Any cable that contains metallic conductors must be properly grounded and bonded.

7. In order to specify a fiber optic cable properly, you need to specify installation specifications as well as environmental specifications.

Multiple Choice
Identify the choice that best completes the statement or answers the question.

8. Cables which contain both multimode and singlemode fibers are called:
    A. Mixed cables
    B. Hybrid cables
    C. Composite cables
    D. XC cables

9. Cables with metallic conductors as well as fiber are called:
    A. Mixed cables
    B. Hybrid cables
    C. Composite cables
    D. XC cables

10. No cable should be installed indoors unless it:
    A. Is UL listed for flame retardancy for NEC
    B. Is colored orange to indicate fiber optics
    C. Is enclosed in innerduct or conduit
    D. The length is printed on the cable jacket

11. ALL outdoor cables are specifically designed to:
    A. Include large numbers of fibers
    B. Be direct buried for ease of installation
    C. Prevent rodent damage to the cable
    D. Prevent moisture damage to the fiber

12. The design of cables of small size with very large fiber counts is usually:
    A. Loose tube
    B. Ribbon
    C. Tight buffer

_13. The design of cables for high pulling tension in outside plant installations is usually:
    A. Loose tube
    B. Ribbon
    C. Tight buffer

_14. The advantage of a breakout cable over a distribution cable is:
    A. The breakout cable has a smaller size and weight
    B. The breakout cable can be installed and terminated without additional hardware for protection of the terminations
    C. Breakout cable costs less than distribution cable
    D. Breakout cable can be used indoors or outdoors

_15. Cables should always be pulled with ______________to prevent damage.
    A. Nylon rope
    B. Kellems grip on the jacket

C. Tension gage
D. The cable's strength members

_16. Loose tube cable requires a _____________to terminate with connectors
A. Splice closure
B. Breakout kit
C. Strain relief
D. Tube stuffer

_17. Armored cable is used in outside plant installations to ___________.
A. Prevent rodent damage
B. Protect from damage from dig-ups
C. Increase pulling tension
D. Conduct lightning strikes

18. The minimum long term bend radius of installed fiber optic cable is usually specified as no less than

________.
A. 12 inches
B. 1 meter
C. 10 times the cable diameter
D. 20 times the cable diameter

_19. Black polyethylene jackets are used on outdoor cables for ____________.
A. Abrasion resistance
B. High tensile load
C. Sunlight and moisture resistance
D. Appearance

20. Match fiber type to jacket color
A. Multimode, 62.5μ
B. Singlemode
C. Multimode, OM3, OM4

1) Aqua
2) Yellow
3) Orange
4) Black
5) Gray

AFTER SECTION 6

From BASICS QUIZ, page 3-9

14. Always keep ____________on connectors when not connected to equipment or being tested.
A. Mating adapters
B. Strain relief boots
C. Sticky tape
D. Dust caps

From TERMINATION AND SPLICING, page 6-34

1. Most singlemode field terminations are made by fusion splicing a factory-made pigtail onto the cable. [T/F]

2. The SC and LC connectors have different size ferrules and cannot me mated. [T/F]

3. The first version of TIA/EIA 568 standard for premises cabling called for the use of which connector?
   A. ST
   B. SC
   C. LC
   D. Any with a FOCIS standard

4. What connector style is now specified in the latest 568 standard (568B.3)?
   A. SC
   B. LC
   C. MT-RJ
   D. Any connector with a FOCIS document

5. Factory terminations, such as used for making patchcords, use what method of attaching the connector to the cable?
   A. Epoxy/polish
   B. Anaerobic adhesive
   C. Pre-polished/splice
   D. Any of the above

6. What is needed to get low loss from a pre-polished/splice connector?
   A. Good stripping technique
   B. Good cleave
   C. Gentle crimp
   D. Proper cable type

7. The difference between a fiber optic connector and a splice is:
   A. Connectors are larger than splices
   B. Connectors are demountable, while splices are permanent
   C. Connectors require adhesives
   D. Splices need expensive tools

8. Which one of the following performance requirements are not shared by connectors and splices:
   A. Low loss
   B. Low back reflection
   C. Repeatability
   D. Durability under repeated mating

9. In singlemode connectors, ____________is as important as low loss.
   A. Ease of field termination
   B. Low reflectance
   C. Low cost
   D. Compatibility with many cable types

_10. Both mechanical splices and pre-polished/splice connectors require a good__________to have low loss.
   A. Field polishing technique
   B. Cleave on the fiber being terminated
   C. Fiber loss
   D. Cable design

11. Physical contact (PC) polish on connectors is designed to reduce _________.
   A. Loss
   B. Reflectance
   C. Loss and reflectance

D. Polishing time

Matching
Identify the following connectors.

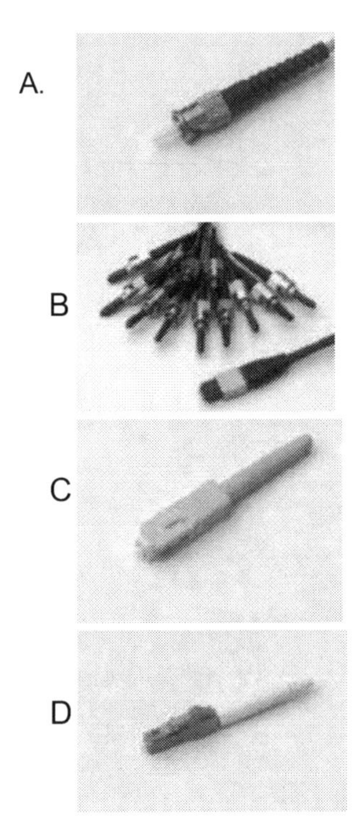
A.
B
C
D

_12. ST

_13. SC

_14. LC

_15. MTP

AFTER SECTION 7

From BASICS QUIZ, page 3-9

___ 9. Concatenation or the joining of two cables in a long outside plant run is almost always done by ____.

A. Mechanical splicing
B. Fusion splicing
C. Field installation of connectors
D. Splicing on pigtailled connectors

___ 11. Premises cables must be rated for ________to meet codes.

A. Pull strength
B. Bend radius
C. Weight in cable trays
D. Fire retardance

From DESIGN 2 QUIZ, page 7-18

__ 1. Fiber optic network designers should have a knowledge of electrical power systems and hardware as well as communications design. [T/F]

___ 2. The first consideration for any network is choosing the proper fiber optic cable type. [T/F]

___ 3. Discussions of which is better – copper, fiber or wireless – are no longer relevant, as fiber is the only choice. [T/F]

___ 4. It may be more cost effective for the fiber optic cabling in many projects to be custom designed and made. [T/F]

__ 5. Testing a fiber optic installation may require testing three times, cable before installation, each segment as installed and a final test of end-to-end loss. [T/F]

__ 6. Fiber optic network designers should have an in-depth knowledge of __________.
A. Fiber optic components and systems
B. Installation processes
C. All applicable standards, codes and any other local regulations
D. All of the above

__ 8. Fiber Optic Network design involves __________.
A. Determining the types of communications systems involved
B. Planning the routes for all cabling or wireless
C. Choosing appropriate cabling and media
D. All of the above

9. Most building management systems use ______________cabling.
A. Fiber optic
B. Coax
C. Structured
D. Proprietary copper

__ 10. Most premises networks today should use _______multimode fiber but backbone cables can contain _____________________________________
fibers for future expansion.
A. OM1, OM3
B. OM1, singlemode
C. OM3, singlemode
D. OM2, OM3

_ 11. _______________of the cable plant is a necessary part of the design and installation process for a fiber optic network that is often overlooked.
A. Planning
B. Documentation
C. CAD-CAM drawing
D. OTDR testing

___ 12. What is the most helpful information you can have when trying to troubleshoot a cabling network for restoration?
A. Phone number of a fiber optic contractor
B. Loss data on each fiber
C. OTDR traces
D. Documentation

13. Metropolitan networks can involve which of the following systems?
A. CCTV surveillance cameras
B. Traffic monitoring
C. Emergency services
D. Educational systems

DESIGN SELF STUDY QUIZ, page 7-18

___ 1. A loss budget is the calculated loss of the cable plant while a power budget is the optical loss tolerable to a communications system. [T/F]

___ 3. Industry standards have high values for component loss compared to typical values expected in normal installations. [T/F]

___ 4. A fiber optic link may have high bit-error rate if ____________.
A. Receiver power is too high
B. Receiver power is too low
C. Receiver power is either too high or too low

___ 5. Loss budgets are used to ensure ___________.
A. The network design will work with the chosen communications equipment
B. Losses of components chosen are appropriate for the cable plant
C. The cable plant tests have a comparison for pass/tail decisions
D. All of the above.

___ 6. When calculating the loss budget, one should choose the component losses using ____.
A. Loss values from industry standards that are always worst case
B. Typical losses that are generally lower than standards
C. Either typical or standard losses as long as it's documented in the design
D. Lowest possible losses so the cable plant loss budget looks better

___ 7. One calculates the contribution of the loss of the fiber to the loss budget by ___________.
A. Looking up the attenuation of the fiber on a manufacturer's data sheet
B. Dividing the length of the fiber by the attenuation
C. Multiplying the length of the fiber by the attenuation coefficient
D. Choosing the best loss possible

___ 9. Connector losses are calculated by adding up all the losses of the connectors, always ____.
A. Including the connectors on each end of the cable plant
B. Including the connectors on each end of the cable plant only if they are connected to a patchcord
C. Excluding the connectors on each end of the cable plant
D. Excluding the connectors on each end of the cable plant if the cable is connected directly to a transceiver

10. A premises cabling link 100 meters long uses multimode fiber (3.0 dB/km @ 850nm) and two

connections in the middle as well as two connectors on the ends (0.50 dB/connector). The calculated loss budget would be

________.

A. 1.30 dB
B. 2.30 dB
C. 3.30 dB
D. 5.00 dB

ANSWER: B
The fiber loss is 3dB/km x 0.1km = 0.3dB plus 4 connectors X 0.5dB = 2.0 dB = 2.3dB[SI 7-68, 7-88]

__ 11. Recalculate the loss budget of the premises cabling link above (100m with 2 connections and connectors on each end) using TIA 568 worst case component losses (fiber at 3.5dB/km and connections at 0.75dB). Then the loss budget now becomes ______.

A. 1.35 dB
B. 2.35 dB
C. 3.35 dB
D. 6.50 dB

From TESTING QUIZ, page 10-41

__ 12. The calculated loss budget can be used as a reference value for testing as well as a check to see if the system will operate over the cable plant being designed, but one needs to allow for a measurement uncertainty of _when setting PASS/FAIL limits.

A. +/- 0.05-0.10 dB
B. +/- 0.2-0.5 dB
C. +/- 0.5-1.0 dB
D. >1 dB

Multiple Response
Identify one or more choices that best complete the statement or answer the question.

13. When calculating a loss budget, one may use values for component loss _________.

A. Specified by industry standards
B. Estimated values of typical installed losses
C. Values specified by the customer
D. Values specified by the component manufacturer

14. When calculating the loss budget of a cable plant, one totals the losses of all the

____

in the link.

A. Fiber attenuation
B. Connections
C. Splices
D. Passive devices like PON splitters
E. Wavelength division multiplexors
F. Transmitter power
G. Receiver sensitivity

From INSTALLATION QUIZ

_ 4. ______________will facilitate installation, allow better planning for upgrades and simplify testing.

A. Good workmanship
B. Low loss connectors

C. Safe workplace procedures
D. Proper documentation

__ 6. The protective gear every VDV installer must always wear is ________.
A. Eye protection
B. Plastic apron
C. Gloves
D. Shoe covers

Testing page 10-41

__ 4. Cable plant loss should be estimated during the ________phase.
A. Design
B. Installation
C. Testing
D. Troubleshooting

__ 12. The total loss of the fiber in the cable plant is calculated by multiplying the attenuation coefficient of the fiber by the ____.
A. Length
B. Number of links
C. Number of connectors
D. Number of splices

AFTER SECTION 8

Design Self Study page 7-18

__ 10. A premises cabling link 100 meters long uses multimode fiber (3.0 dB/km @ 850nm) and two connections in the middle as well as two connectors on the ends (0.50 dB/connector). The calculated loss budget would be ________.
E. 1.30 dB
F. 2.30 dB
G. 3.30 dB
H. 5.00 dB

__ 11. Recalculate the loss budget of the premises cabling link above (100m with 2 connections and connectors on each end) using TIA 568 worst case component losses (fiber at 3.5dB/km and connections at 0.75dB). Then the loss budget now becomes ______.
E. 1.35 dB
F. 2.35 dB
G. 3.35 dB
H. 6.50 dB

AFTER SECTION 9

From INSTALLATION QUIZ, page 9-16

_ 5. Outside plant cabling can be installed by __________.
A. Pulling in underground in conduit
B. Direct burial
C. Aerial suspension

D. All of the above

__ 2. All metal components of the cabling system installed in a equipment or telecom room must be grounded and bonded. [T/F]

__ 3. When upgrading cables in a telecom closet, old, abandoned cables can be cut back to the wall and left in place, as long as the firestopping is not disturbed. [T/F]

__ 7. The fiberglass rod inside many fiber optic cables is for
A. Increasing the pulling tension
B. Limit bend radius to preventing kinking
C. Winding the fibers around
D. Tying to messenger cables

Identify the choice that best completes the statement or answers the question.

__ 8. To prevent the cable from twisting when pulling it
A. Use a swivel eye
B. Pull with braided rope
C. Spin the cable off the spool
D. Lubricate the cable

__ 9. On long pulls, at intermediate points, why do you lay the cable in a "figure 8'?
A. Keep it from getting tangled with the pull rope
B. Make it easier to spray on lubricant
C. Keep workers from walking on it
D. Prevent it from twisting

__ 10. Under pulling tension, the bend radius should not be less than
A. 5 times the cable diameter
B. 10 times the cable diameter
C. 20 times the cable diameter
D. 50 times the cable diameter

_ 12. The protective gear every VDV installer must always wear is __________.
A. Eye protection
B. Plastic apron
C. Gloves
D. Shoe covers

__ 13. Vertical cable runs should preferably be installed by __________.
A. Pulling slowly and carefully by hand
B. Calibrated pulling machines
C. Pulling one floor at a time
D. Dropping from above rather than pulling up

__ 14. Cable ties used on fiber optic cables ___________.
A. Should be tightened firmly to prevent cable movement
B. Can be used to hang cables from J-hooks or cable trays
C. Should be rated for the weight of the cables
D. Can harm cables if too tight, so they should be hand-tightened
E.

AFTER SECTION 10

__ 10. Which fiber optic test instrument uses backscattered light for measurements?
A. OLTS
B. OTDR
C. VFL
D. Tracer

FROM testing quiz, PAGE 10-41

1. Cables tested with an OTDR do not require insertion loss testing with a source and meter or OLTS. [T/F]

__ 2. Connectors at each end of the cable plant should not be counted when calculating the cable plant loss. [T/F]

__ 3. The OTDR should never be used without a "launch cable" which is also called a "pulse suppressor." [T/F]

__ 5. The standard method of testing installed multimode cables in a cable plant is described in:
A. FOTP-34
B. ISO 11801
C. FOTP-57
D. OFSTP-14

__ 6. What test instrument(s) are used for insertion loss testing.
A. OLTS or light source and power meter
B. VFL
C. OTDR

__ 7. Multimode graded-index glass fiber optic cables are tested with _____sources at ____and ____________ wavelengths.
A. LED, 650, 850 nm
B. LED, 850, 1300 nm
C. Laser, 980, 1400 nm
D. Laser, 1310, 1550 nm

__ 8. What type of source is used for testing singlemode fibers?
A. LED
B. VCSEL
C. Laser

__ 9. How many methods are included in standards for setting the "0 dB' reference for loss testing?
A. One
B. Two
C. Three
D. Four

__ 10. Which reference method is required for TIA 568?
A. Once cable reference
B. Two cable reference
C. Three cable reference
D. Any method as long as it is documented

__ 11. Reference cables must match the __________of the cables being tested.
A. Fiber size and type

B. Fiber size and connector type
C. Connector type
D. Fiber size and loss specification

__ 13. The principle of operation of OTDRs is similar to _________.
A. Power meters and sources
B. Radar
C. Mirrors
D. Lenses

__ 14. OTDRs are used in outside plant cables to ____________.
A. Verify splice loss
B. Measure length
C. Find faults
D. All of the above

__ 15. In premises applications, OTDRs are limited in usefulness by their ________.
A. Output power
B. Distance capability
C. Distance Resolution
D. Software

WEB ADDRESSES OF FOA CFOT PRACTICE QUIZZES

- https://fiberu.org/basic/index.html
- https://www.thefoa.org/tech/ref/basic/quiz/jargon/jargon.htm
- https://www.thefoa.org/tech/ref/basic/quiz/link/link.htm
- https://www.thefoa.org/tech/ref/basic/quiz/basic/basic.htm
- https://www.thefoa.org/tech/ref/basic/quiz/networks/networks.htm
- https://www.thefoa.org/tech/ref/basic/quiz/fiber/fiber.htm
- https://www.thefoa.org/tech/ref/basic/quiz/cables/cables.htm
- https://www.thefoa.org/tech/ref/basic/quiz/term/term.htm
- https://www.thefoa.org/tech/ref/basic/quiz/design/design.htm
- https://fiberu.org/Design/Quizzes/LP4quiz.htm
- https://www.thefoa.org/tech/ref/basic/quiz/install/install.htm
- https://www.thefoa.org/tech/ref/basic/quiz/test/test.htm

f: Appendix All quizzes 7-30-24

Pearson Technologies Inc.

4671 Hickory Bend Dr Acworth GA 30102 770-490-9991 www.ptnowire.com fiberguru@ptnowire.com

30 July 2024

Handout List for PFOI+NETW3523

Page

List of Handouts, pfoi+3523.docx7/30/24 9:26 AM

Handout 2
3523-TEAM ASSIGNMENTS FOR END PREPARATION FOR 24 TRAINEES

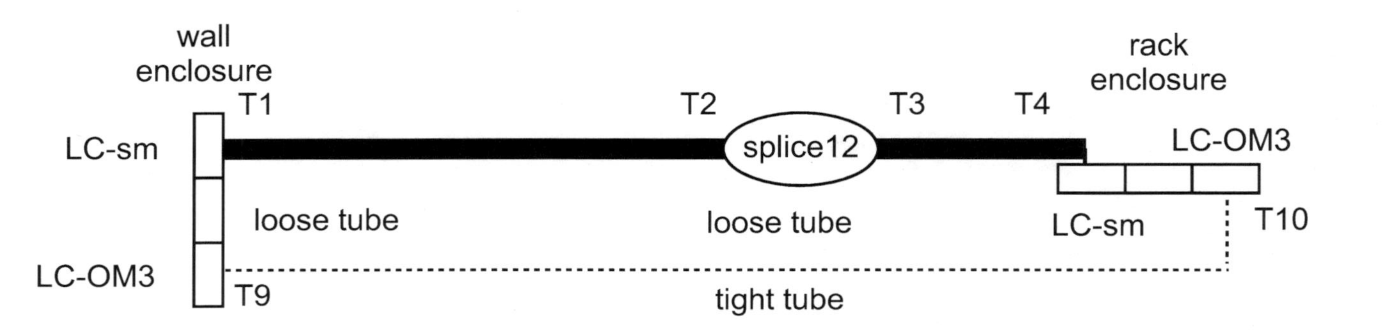

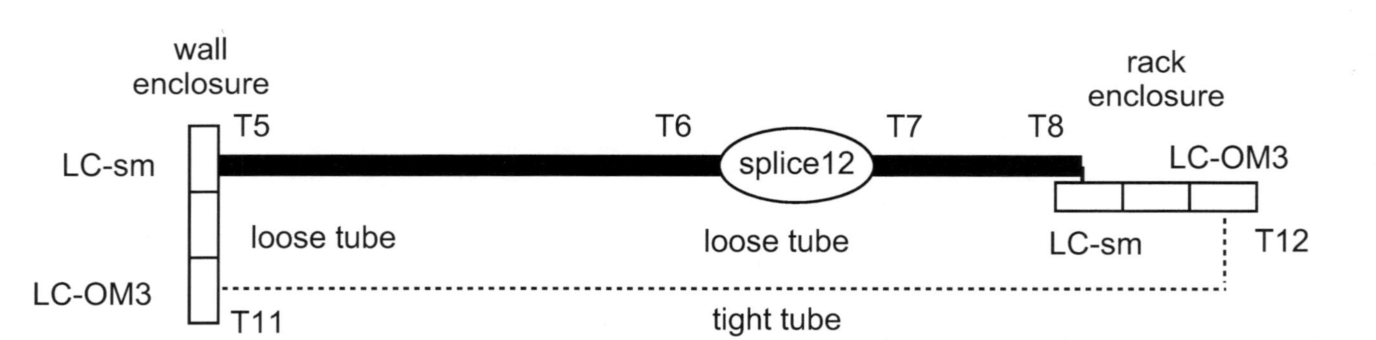

PLAN

Instructor to assign fiber color and team number (Tx) to each trainee at beginning; trainee is to splice only that color; splice fiber as indicatedboth mid span and end enclosures

Except T9-T12, all teams require cable prep kits with blue slitter. For outdoor enclosure, instructor to trim fibers to decreasing length as fiber number increases. See Form 5?.

first splicing round mid span splicing, in red with 1 splice/trainee
simultaneously, 1-2 splicers at indoor enclosures; with 1 splice /trainee;
each trainee splices same fiber color in both end enclosures
as splicers finish at mid span, move to end enclosures with 1 splice/trainee

use higher number fibers for demo; instructor to splice 2 fibers in demo;
trainees splice number indicated

tight tube end prep done by first teams to finish loose tube prep

Hand Out 2b
3523-TEAM ASSIGNMENTS FOR END PREPARATION FOR 15 TRAINEES

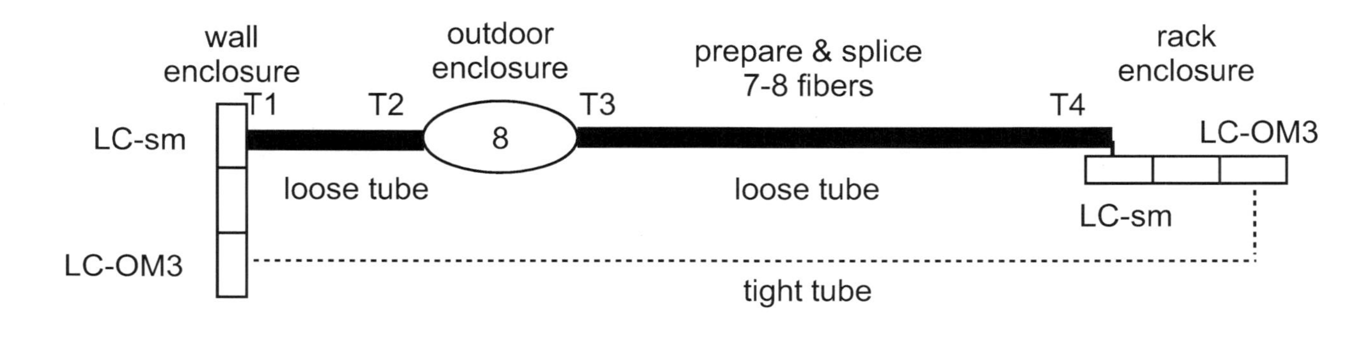

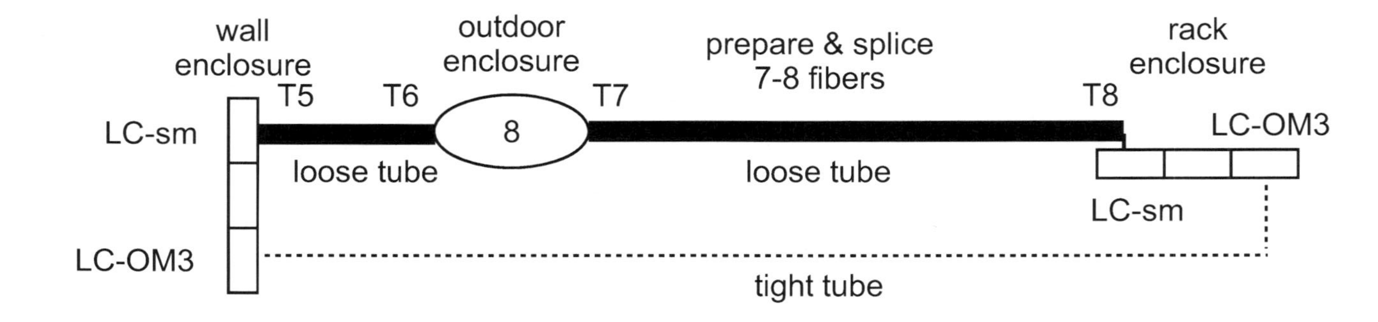

PLAN

instructor to assign fiber color to each trainee at beginning; trainee is to splice only that color; splice fibers 1-8 at both mid span and end enclosures

in outdoor enclosure, instructor to trim fibers to decreasing length as fiber number increases

first splicing round mid span splicing, in red with 1 splice/trainee
simultaneously, 1-2 splicers at indoor enclosures; with 1 splice /trainee;
each trainee splices same fiber color in both end enclosures
as splicers finish at mid span,move to end enclosures with 1 splice/trainee

use higher number fibers for demo; instructor to splice 2 fibers in demo;
trainees splice 8 fibers

simultaneously VFL 6 fibers in one direction; 6 fibers in opposite;

Hand Out 2c
3523-TEAM ASSIGNMENTS FOR END PREPARATION FOR 10 TRAINEES

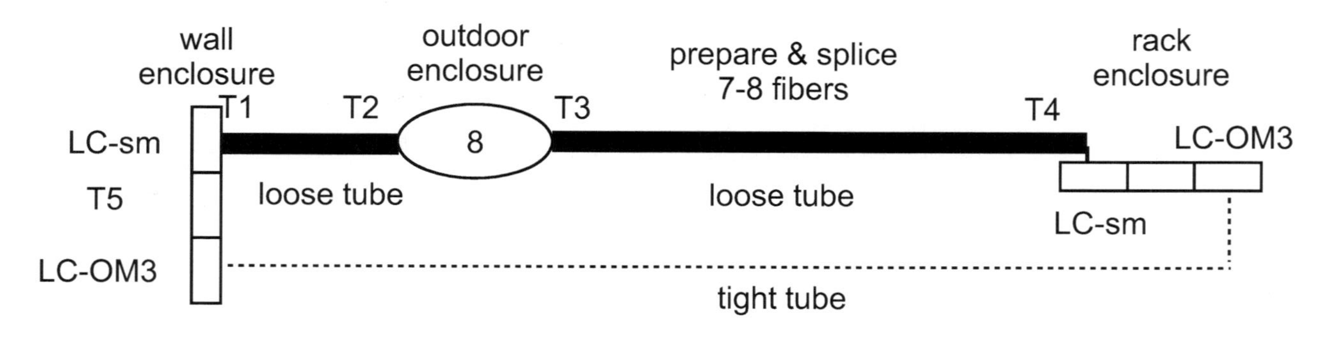

PLAN

instructor to assign fiber color to each trainee at beginning; trainee is to splice only that color; splice fibers 1-8 at both mid span and end enclosures

in outdoor enclosure, instructor to trim fibers to decreasing length as fiber number increases

first splicing round mid span splicing, in red with 1 splice/trainee
simultaneously, 1-2 splicers at indoor enclosures; with 1 splice /trainee;
each trainee splices same fiber color in both end enclosures
as splicers finish at mid span,move to end enclosures with 1 splice/trainee

use higher number fibers for demo; instructor to splice 2 fibers in demo;
trainees splice 8 fibers

simultaneously VFL 6 fibers in one direction; 6 fibers in opposite;

Handout 3
3523-SPLICER STARTING LOCATIONS FOR 24 TRAINEES

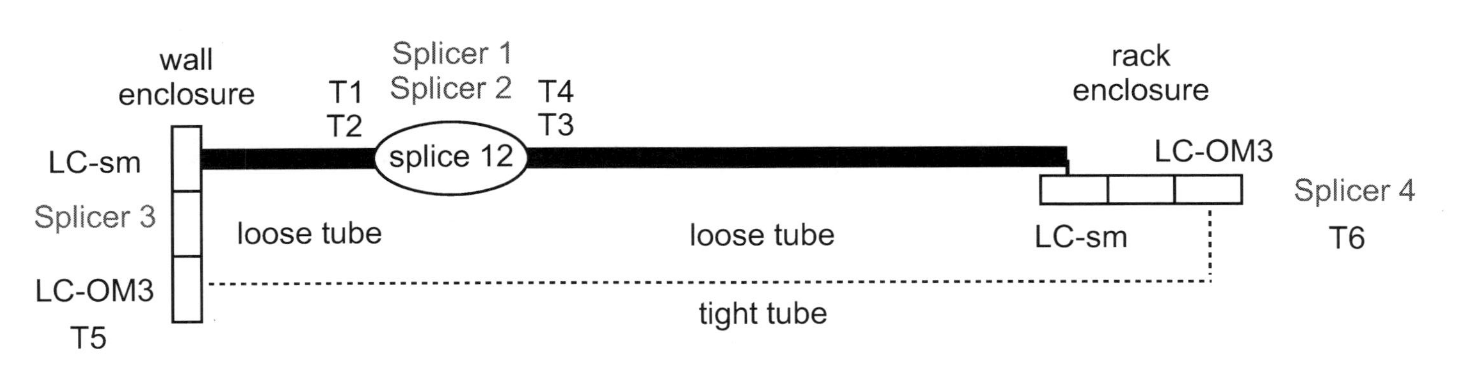

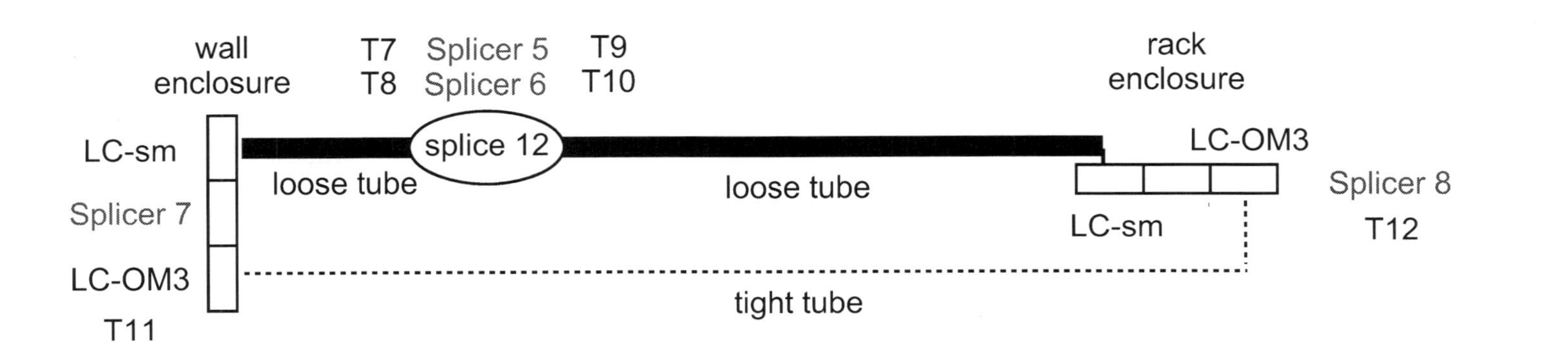

PLAN

instructor to assign fiber color to each trainee at beginning; trainee is to splice only that color; splice fibers 1-8 at both mid span and end enclosures

in outdoor enclosure, instructor to trim fibers to decreasing length as fiber number increases

first splicing round mid span splicing, in red with 1 splice/trainee
simultaneously, 1-2 splicers at indoor enclosures; with 1 splice /trainee;
each trainee splices same fiber color in both end enclosures
as splicers finish at mid span, move to end enclosures with 1 splice/trainee

use higher number fibers for demo; instructor to splice 2 fibers in demo; trainees splice 8 fibers

Handout 3b
3523-SPLICER STARTING LOCATIONS FOR 15 TRAINEES

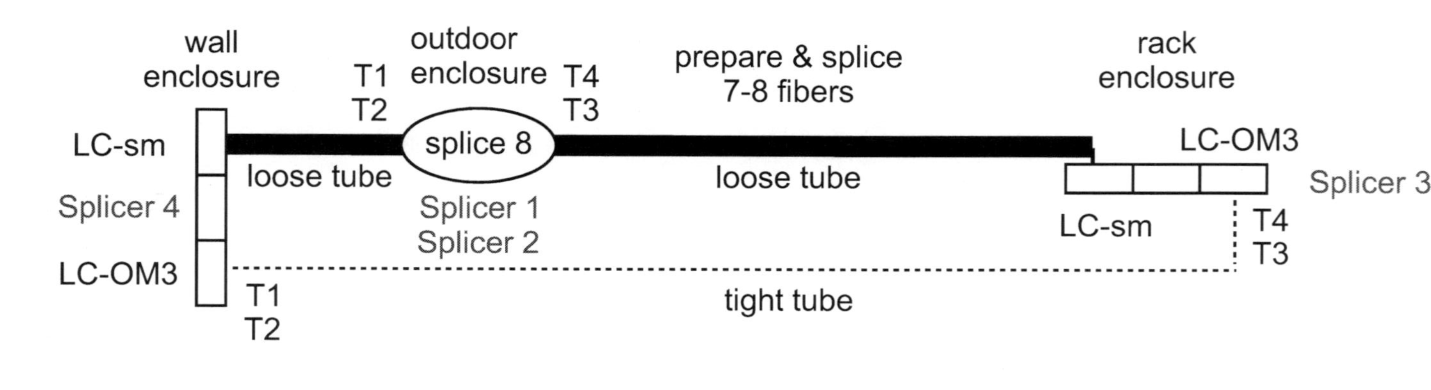

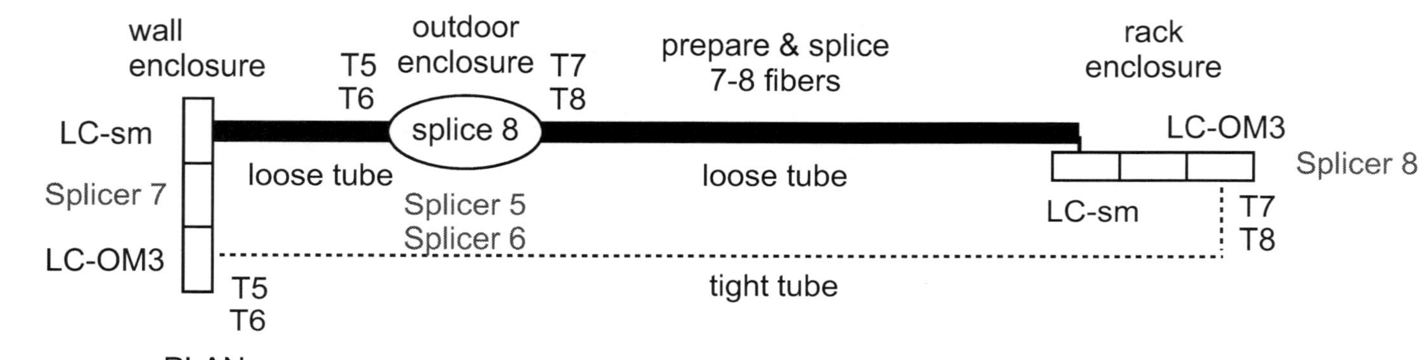

PLAN

instructor to assign fiber color to each trainee at beginning; trainee is to splice only that color; splice fibers 1-8 at both mid span and end enclosures

in outdoor enclosure, instructor to trim fibers to decreasing length as fiber number increases

first splicing round mid span splicing, in red with 1 splice/trainee
simultaneously, 1-2 splicers at indoor enclosures; with 1 splice /trainee;
each trainee splices same fiber color in both end enclosures
as splicers finish at mid span,move to end enclosures with 1 splice/trainee

use higher number fibers for demo; instructor to splice 2 fibers in demo;
trainees splice 8 fibers

simultaneously VFL 6 fibers in one direction; 6 fibers in opposite;

Handout 3c
3523-SPLICER STARTING LOCATIONS FOR 10 TRAINEES

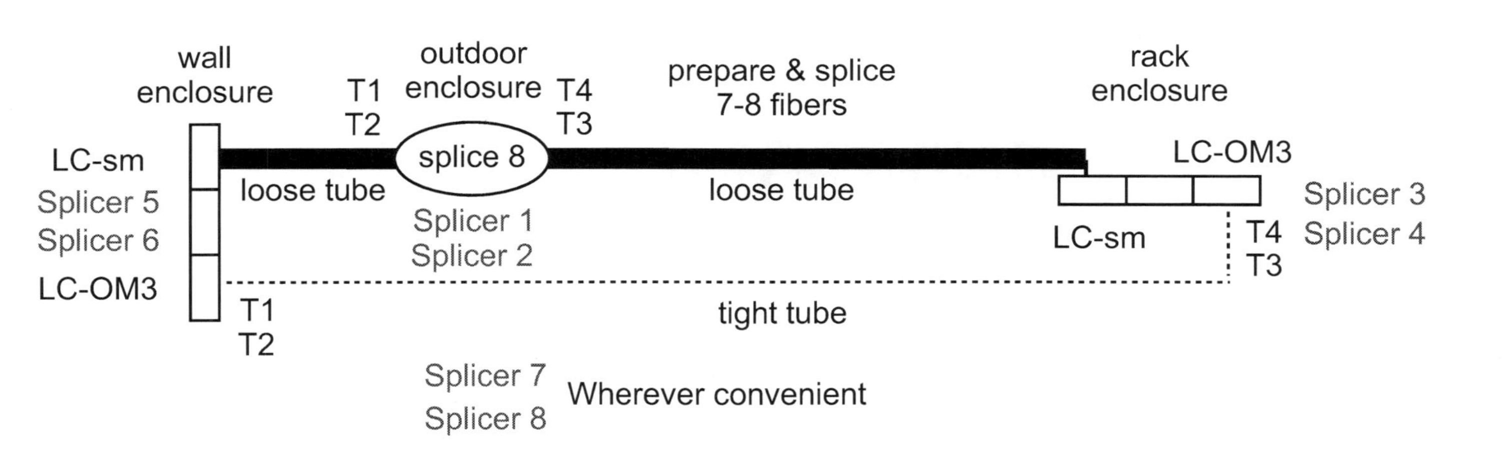

PLAN

instructor to assign fiber color to each trainee at beginning; trainee is to splice only that color; splice fibers 1-8 at both mid span and end enclosures

in outdoor enclosure, instructor to trim fibers to decreasing length as fiber number increases

first splicing round mid span splicing, in red with 1 splice/trainee
simultaneously, 1-2 splicers at indoor enclosures; with 1 splice /trainee;
each trainee splices same fiber color in both end enclosures
as splicers finish at mid span,move to end enclosures with 1 splice/trainee

use higher number fibers for demo; instructor to splice 2 fibers in demo;
trainees splice 8 fibers

simultaneously VFL 6 fibers in one direction; 6 fibers in opposite;

4671 Hickory Bend Dr Acworth GA 30102 770-490-9991 www.ptnowire.com fiberguru@ptnowire.com

Handout 5
Fiber Length Change, and Verification of Proper Fiber Selection

Installation of Cables into Enclosure

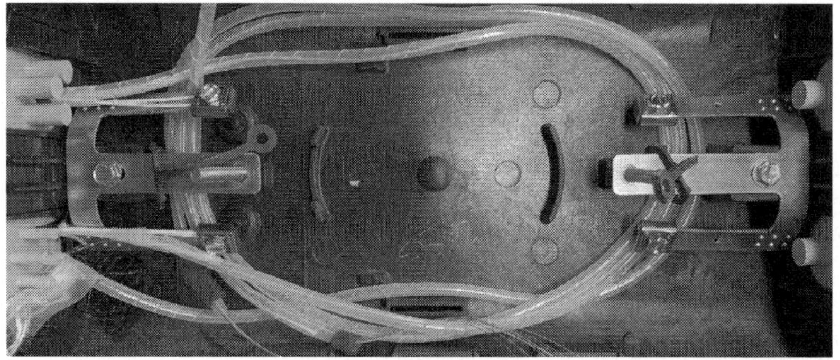

Ignore third spiral wrap with 2 fibers at lower left. Note strength members under clamp at left end of photograph. After spiral wrap installed on both cables, coil wrap as shown.

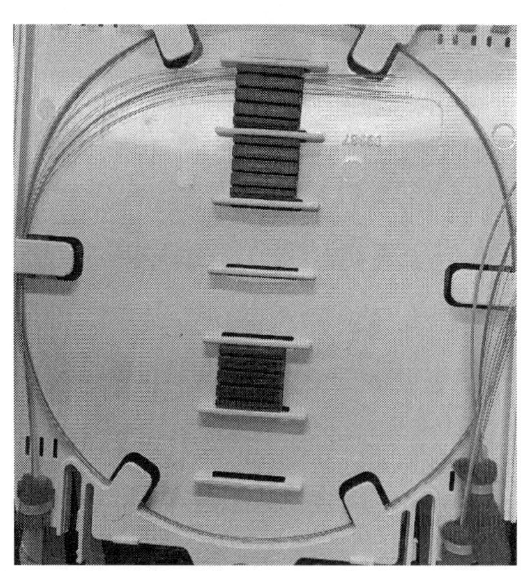

Attach spiral wraps to splice tray with cable tie. Latch for tie is on underside of tray.

Fiber Length Change

Before splicing, route the fibers in the tray to the first/top splice holder. Cut all fibers at a length so that the ends extend 1/4"-3/8" beyond the splice holder, as shown below. Shorten fibers 5-12 by ¾". Shorten fibers 9-12 by ¾".

4671 Hickory Bend Dr Acworth GA 30102 770-490-9991 www.ptnowire.com fiberguru@ptnowire.com

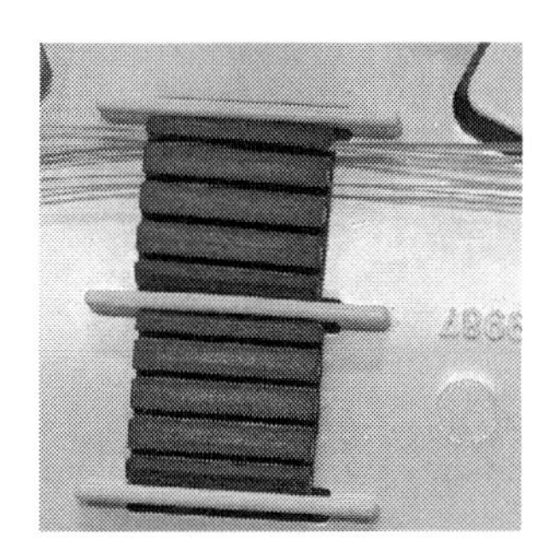

Appearance Prior to First Length Cut

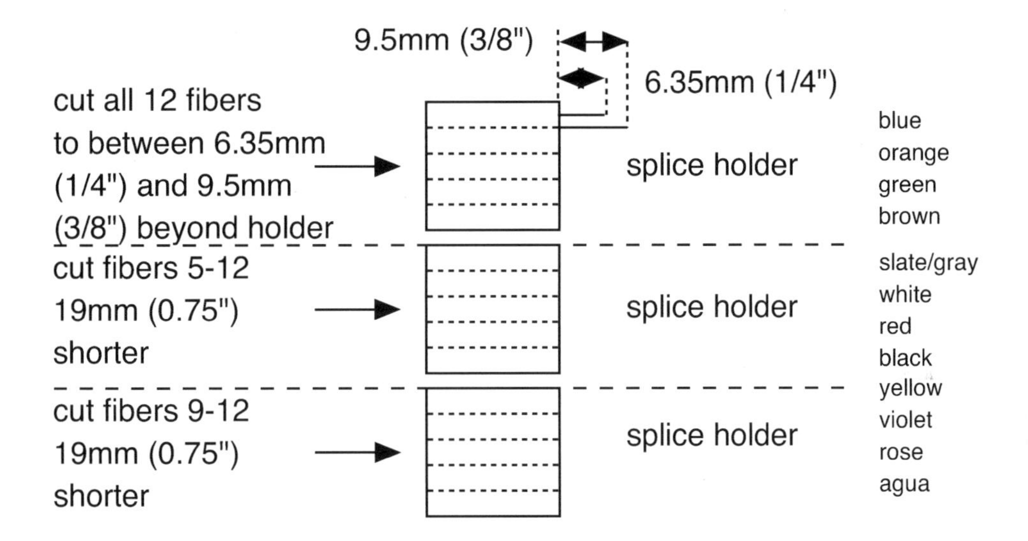

Verification of Proper Fiber Selection

After selecting fibers to splice, both team members verify fibers selected as being correct..

Fiber Selection Verification+length delta.docx
1/2/24 7:17 AM

Handout 6
Splice Tray Circuits

Fibers Enter Opposite Sides

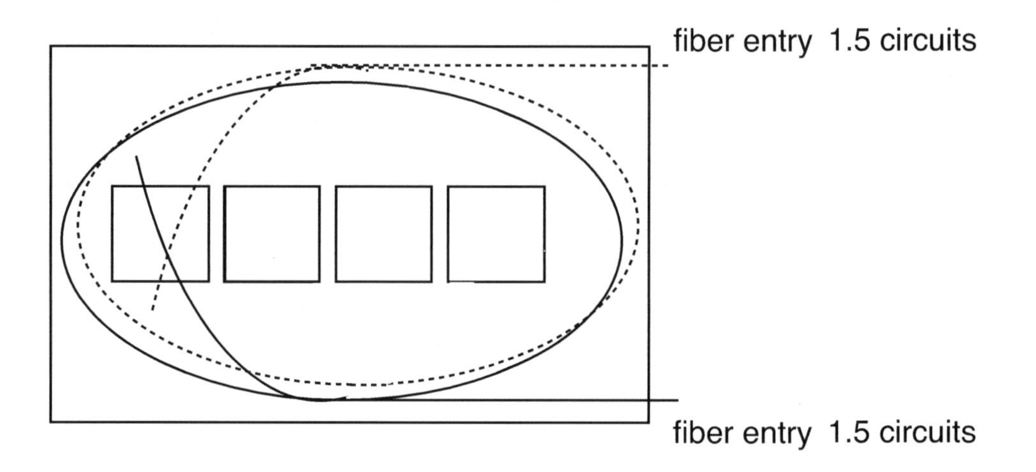

Fibers Enter Same Side

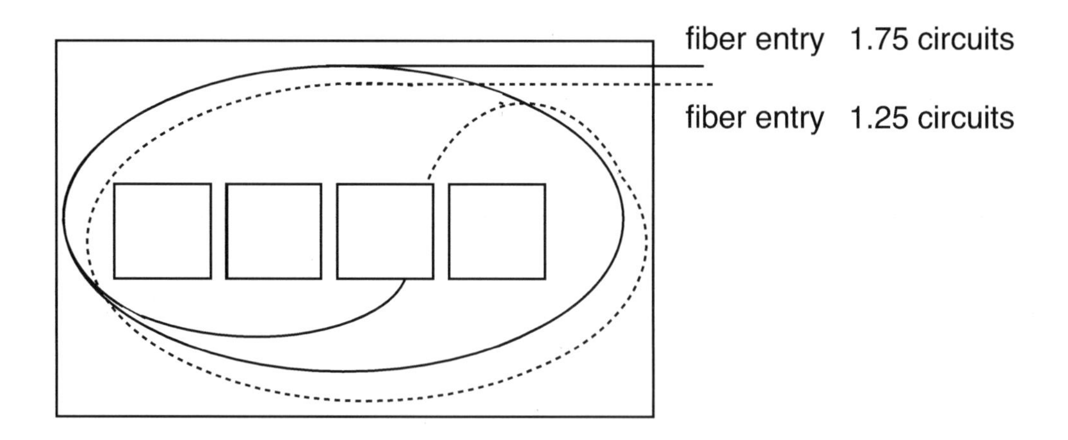

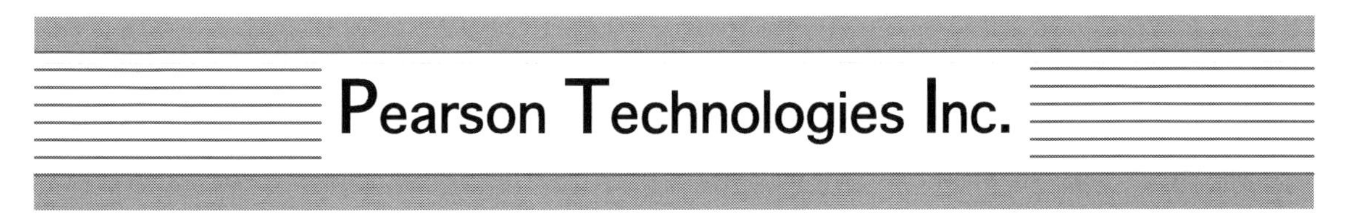

4671 Hickory Bend Dr Acworth GA 30102 770-490-9991 www.ptnowire.com fiberguru@ptnowire.com

Handout 7

SUPPPLEMENTAL INSTRUCTIONS: INNO FUSION SPLICERS

1. FOR ALL INNOs
   Check splice menu setting for fiber type. For most programs, the setting is either "singlemode auto" or "657 to 657".

   STRIP LENGTH: strip buffer tube and primary coating to 1.375" ±0.128" as shown below.

   Goal ______________

   Minimum ______________

   Maximum ______________

2. For SOCs: Cleave length is 10mm.

3. FOR SOCs EXCEPT WITH IFS55
   Check heater menu setting. The setting is either SOC or 31mm.

4. FOR SOCs WITH INNO IFS55
   Use external oven.

5. FOR SPLICING OTHER THAN SOCs (250μ to 250μ or 900μ; 250μ to 250μ)
   Check heater menu setting.
   The splice cover length is 40mm for program NETW3523.
   The splice cover length is 40mm for FTTH programs.

VFL Evaluations

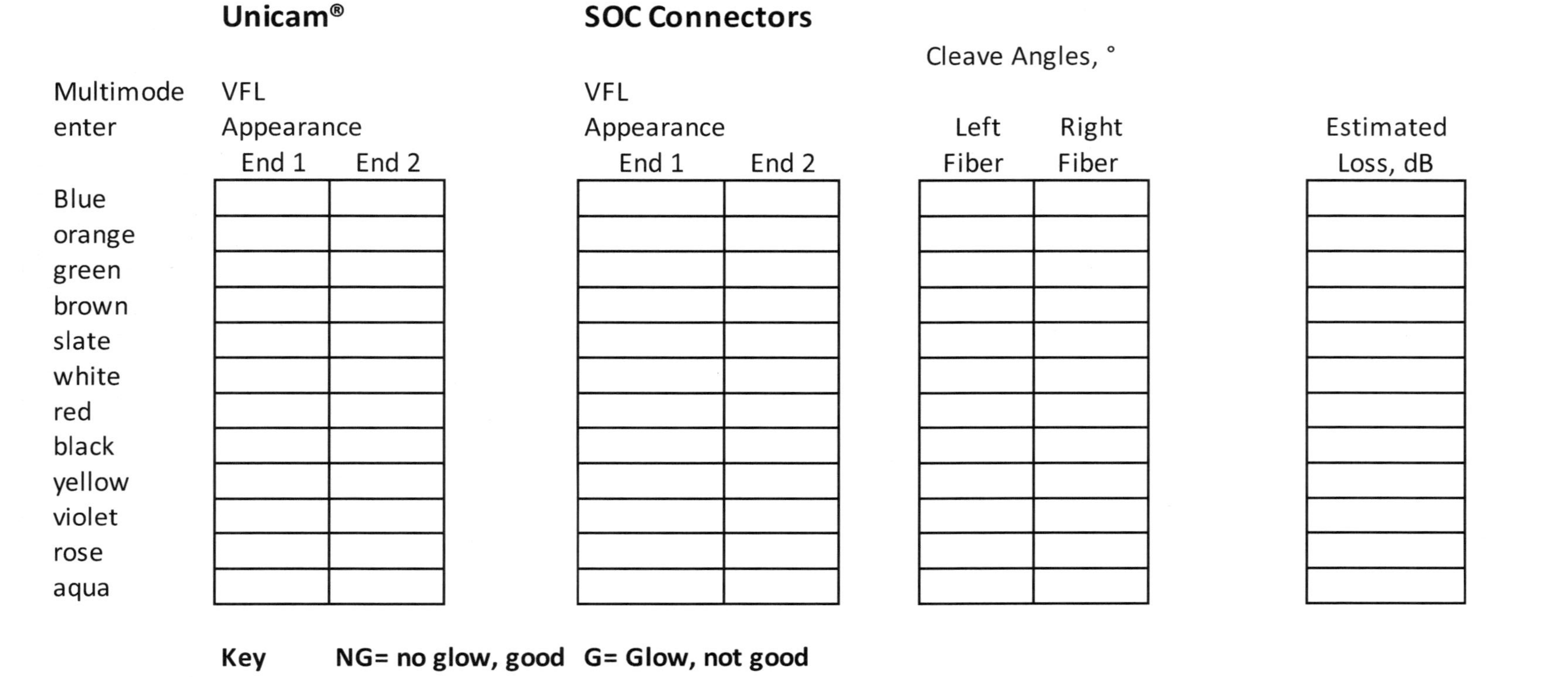

Multimode enter	Unicam® VFL Appearance End 1	Unicam® VFL Appearance End 2	SOC Connectors VFL Appearance End 1	SOC Connectors VFL Appearance End 2	Cleave Angles, ° Left Fiber	Cleave Angles, ° Right Fiber	Estimated Loss, dB
Blue							
orange							
green							
brown							
slate							
white							
red							
black							
yellow							
violet							
rose							
aqua							

Key **NG= no glow, good** **G= Glow, not good**

Hand Out 9

Unicam® 850nm Insertion Loss Measurement

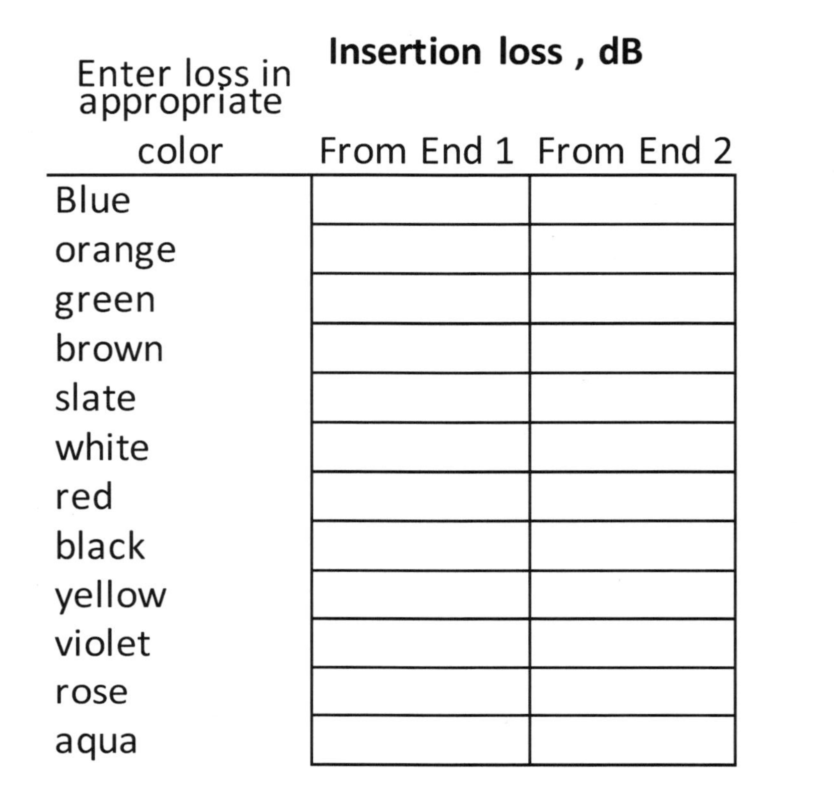

Enter loss in appropriate color	Insertion loss , dB	
	From End 1	From End 2
Blue		
orange		
green		
brown		
slate		
white		
red		
black		
yellow		
violet		
rose		
aqua		

Name: ______________

Date: _________

Handout 10

Insertion Loss, 1310nm

Link 1

from rack end	team 1	team 2	team 3	team 4	team 5	team 6
1						
2						
3						
4						
5						
6						
7						
8						
9						
10						
11						
12						

from opposite end	team 1	team 2	team 3	team 4	team 5	team 6
1						
2						
3						
4						
5						
6						
7						
8						
9						
10						
11						
12						

Link 2

from rack end	team 7	team 8	team 9	team 10	team 11	team 12
1						
2						
3						
4						
5						
6						
7						
8						
9						
10						
11						
12						

from opposite end	team 7	team 8	team 9	team 10	team 11	team 12
1						
2						
3						
4						
5						
6						
7						
8						
9						
10						
11						
12						

Conclusion: Are losses in opposite directions exactly the same? | Yes | No |

Names: ________________ **Handout 11** Date: _________

Range Testing, 1310nm

RANGE TESTING, 1310nm

results from rack Measurement	Link 1 fiber 1 team 1	 fiber 2 team 2	 fiber 3 team 3	 fiber 4 team 4	 fiber 5 team 5	 fiber 6 team 6
1						
2						
3						
4						
5						
6						
largest value						
smallest value						
Difference=						

from rack Measurement	Link 2 fiber 1 team 7	 fiber 2 team 8	 fiber 3 team 9	 fiber 4 team 10	 fiber 5 team 11	 fiber 6 team 12
1						
2						
3						
4						
5						
6						
largest value						
smallest value						
Difference=						

Range= Larger of 12 differences

Range=						

Instructions

Each term record its range and enter ranges from other teams.

Conclusion		
Conclusion: Are all loss measurements exactly the same?	Yes	No
Conclusion: Are all range measurements exactly the same?	Yes	No

Handout 12
OTDR TEST RESULTS

Instructions: Each team to enter its own data and collect data from other teams.

LINK 1

Team		Fiber	**launch cable**		**segment 1**			**segment 2**			attenuation Uniform?	
			length m	connector loss, dB	length m	attenuation rate, dB/km	splice loss @end, dB	length m	attenuation rate, dB/km	connector loss, dB	**segment 1** Yes/No	**segment 2** Yes/No
1	forward	blue										
2	forward	orange										
3	forward	green										
4	forward	brown										
5	forward	slate										
6	forward	white										
7	forward	red										
8	forward	black										
9	forward	yellow										
10	forward	violet										
11	forward	rose										
12	forward	aqua										
1	reverse	blue										
2	reverse	orange										
3	reverse	green										
4	reverse	brown										
5	reverse	slate										
6	reverse	white										
7	reverse	red										
8	reverse	black										
9	reverse	yellow										
10	reverse	violet										
11	reverse	rose										
12	reverse	aqua										

Handout 12
OTDR TEST RESULTS

Instructions: Each team to enter its own data and collect data from other teams.

LINK 2

Team		Fiber	**launch cable**		**segment 1**			**segment 2**			attenuation Uniform?	
			length m	connector loss, dB	length m	attenuation rate, dB/km	splice loss @end, dB	length m	attenuation rate, dB/km	connector loss, dB	**segment 1** Yes/No	**segment 2** Yes/No
1	forward	blue										
2	forward	orange										
3	forward	green										
4	forward	brown										
5	forward	slate										
6	forward	white										
7	forward	red										
8	forward	black										
9	forward	yellow										
10	forward	violet										
11	forward	rose										
12	forward	aqua										
1	reverse	blue										
2	reverse	orange										
3	reverse	green										
4	reverse	brown										
5	reverse	slate										
6	reverse	white										
7	reverse	red										
8	reverse	black										
9	reverse	yellow										
10	reverse	violet										
11	reverse	rose										
12	reverse	aqua										

Handout 13

Names: ________________

Date: _________

Microscopic Inspection

MICROSCOPIC INSPECTION, FIRST

G=GOOD NG=NOT GOOD

Link 1

results from rack	team 1	team 2	team 3	team 4	team 5	team 6
1						
2						
3						
4						
5						
6						
7						
8						
9						
10						
11						
12						

Link 2

results from rack	team 7	team 8	team 9	team 10	team 11	team 12
1						
2						
3						
4						
5						
6						
7						
8						
9						
10						
11						
12						

MICROSCOPIC INSPECTION, AFTER CLEANING

G=GOOD NG=NOT GOOD

results from rack	team 1	team 2	team 3	team 4	team 5	team 6
1						
2						
3						
4						
5						
6						
7						
8						
9						
10						
11						
12						

results from rack	team 7	team 8	team 9	team 10	team 11	team 12
1						
2						
3						
4						
5						
6						
7						
8						
9						
10 11						
12						

Names: ________________ **Handout 13** Date: __________

Microscopic Inspection

MICROSCOPIC INSPECTION, FIRST

G=GOOD NG=NOT GOOD

Link 1

results from wall	team 1	team 2	team 3	team 4	team 5	team 6
1						
2						
3						
4						
5						
6						
7						
8						
9						
10						
11						
12						

Link 2

results from wall	team 7	team 8	team 9	team 10	team 11	team 12
1						
2						
3						
4						
5						
6						
7						
8						
9						
10						
11						
12						

MICROSCOPIC INSPECTION, AFTER CLEANING

G=GOOD NG=NOT GOOD

results from wall	team 1	team 2	team 3	team 4	team 5	team 6
1						
2						
3						
4						
5						
6						
7						
8						
9						
10						
11						
12						

results from wall	team 7	team 8	team 9	team 10	team 11	team 12
1						
2						
3						
4						
5						
6						
7						
8						
9						
10						
11						
12						

Handout 14

Acceptance Values

For Exercise 19.2 in Professional Fiber Optic Installation, v10

MAXIMUM LOSS **1310nm**

	Max. dB/km		# km		dB
Cable loss=		x		=	
	Max. dB/pair		# pairs		
Connector loss=		x		=	
	Max. dB/splice		# splices		
Splice loss=		x		=	
	Max. dB/splitter		# splitters		
Splitter loss=		x		=	
			Total loss	=	

TYPICAL LOSS **1310nm**

	Typ. dB/km		# km		dB
Cable loss=		x		=	
	Typ. dB/pair		# pairs		
Connector loss=		x		=	
	Typ. dB/splice		# splices		
Splice loss=		x		=	
	Typ. dB/splitter		# splitters		
Splitter loss=		x		=	
			Total loss	=	

Insertion loss acceptance value = ______ dB

OTDR Acceptance Values

1310	Max.	
Attenuation rate	0.5	dB/km
Connector loss	0.75	dB/connection
Splice loss	0.15	dB/splice

1310	Typ.	
Attenuation rate	0.35	dB/km
Connector loss	0.15	dB/connection
Splice loss	0.05	dB/splice

Acceptance Values		
Attenuation rate		dB/km
Connector loss		dB/connection
Splice loss		dB/splice

Handout 15

Acceptance Values

INSTRUCTION: Use values from OTDR testing

Link 1		***check Link #***
Link 2		

MAXIMUM LOSS **1310nm**

	Max. dB/km		# km		dB
Cable loss=		x		=	
	Max. dB/pair		# pairs		
Connector loss=		x		=	
	Max. dB/splice		# splices		
Splice loss=		x		=	
	Max. dB/splitter		# splitters		
Splitter loss=		x		=	
			Total loss	=	

TYPICAL LOSS **1310nm**

	Typ. dB/km		# km		dB
Cable loss=		x		=	
	Typ. dB/pair		# pairs		
Connector loss=		x		=	
	Typ. dB/splice		# splices		
Splice loss=		x		=	
	Typ. dB/splitter		# splitters		
Splitter loss=		x		=	
			Total loss	=	

Insertion loss acceptance value = ______ dB

OTDR Acceptance Values

1310	Max.	
Attenuation rate	0.5	dB/km
Connector loss	0.75	dB/connection
Splice loss	0.15	dB/splice

1310	Typ.	
Attenuation rate	0.35	dB/km
Connector loss	0.15	dB/connection
Splice loss	0.05	dB/splice

Acceptance Values		
Attenuation rate		dB/km
Connector loss		dB/connection
Splice loss		dB/splice

Made in the USA
Columbia, SC
06 January 2025